energy
efficient
site design

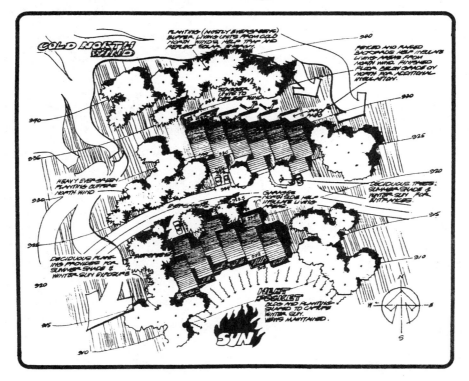

edited by
Gary O. Robinette

based on studies undertaken by the
Reimann, Buechner, Crandall Partnership
for the
Center for Landscape Architectural Education and Research

VAN NOSTRAND REINHOLD COMPANY
NEW YORK   CINCINNATI   TORONTO   LONDON   MELBOURNE

Paperback edition originally published by Environment Design Press.
Paperback edition copyright © 1981 by Environment Design Press.

Copyright © 1983 by Van Nostrand Reinhold Company Inc.

Library of Congress Catalog Card Number: 83-1073
ISBN: 0-442-22338-2

Manufactured in the United States of America

Published by Van Nostrand Reinhold Company Inc.
135 West 50th Street, New York, N.Y. 10020

Van Nostrand Reinhold Publishing
1410 Birchmount Road
Scarborough, Ontario M1P 2E7, Canada

Van Nostrand Reinhold
480 Latrobe Street
Melbourne, Victoria 3000, Australia

Van Nostrand Reinhold Company Limited
Molly Millars Lane
Wokingham, Berkshire, England

15 14 13 12 11 10 9 8 7 6 5 4 3 2 1

Library of Congress Cataloging in Publication Data
Main entry under title:

Energy efficient site design.

   "Based on studies undertaken by the Reimann,
Buechner, Crandall Partnership."
   Bibliography: p.
   1. Architecture and energy conservation—United
States. 2. Architecture and climate—United States.
3. Building sites—United States—Planning. 4. Archi-
tecture and energy conservation—Canada. 5. Archi-
tecture and climate—Canada. 6. Building sites—Canada
   Planning. I. Robinette, Gary O. II. Reimann,
Buechner, Crandall Partnership.
NA2542.3.E48   1983   720'.47   83-1073
ISBN 0-442-22338-2

abstract

There is a traditionally recognized and scientifically calculated comfort zone for the human body. Throughout history man has either moved to a climate which is either closer to that comfort zone or has altered the environment to bring it close to that comfort zone. In the past man has built and designed to take advantage of the natural heating and cooling potential of the sun or wind. With the advent of modern technology man has been able to create, with the expenditure of vast amounts of energy, nearly ideal interior environments, regardless of the outdoor climatic conditions.

At a time of concern for energy supplies and costs it is essential to examine the energy conserving potential of once again using natural energy systems and patterns which exist on any site on which a building is to be or has been placed. This is a state-of-the-art study examining the existing research and existing applications of land planning and design for energy conservation purposes. It is organized around the basic principles of natural heating and cooling factors and the impact of site elements on human comfort, the regional adaptation of these principles, the steps in the site planning and design process and what energy conservation options are available in each of the steps. This is supplemented by a bibliography of additional references.

The principles of site planning and design for energy conservation deal with the impact of the sun and wind on natural elements and conversely the ways in which landforms, water and vegetation affect the impact of the sun and wind on limited sections of the earth. For instance, the sun is able to naturally warm certain slopes more than others, certain surface materials naturally accentuate the warmth of the sun, while certain naturally occurring elements, such as vegetation, block and control the sun. The wind moving over the earth's surface has certain natural directions, patterns and characteristics. By recognizing these and by introducing or removing impediments it is possible to modify these patterns to provide natural ventilation and to conserve energy.

Four commonly accepted climatic regional divisions were utilized to illustrate the differing applications of these principles in the temperate, hot-humid, hot-arid and the cool regions of the continental United States. The typical site planning or design process was superimposed over each depending upon the precise decision made in each step in each region.

introduction

Almost all human cultures in modern society build with forms and in places with only a minimum of concern for the natural processes of the sun or wind which can heat or cool naturally and inexpensively. Modern technology enables human beings to make any area, whether in the desert or in the arctic, liveable and comfortable through the use of artificial and energy-intensive heating, air conditioning or humidity control. All of this is done, however, through the expenditure of great amounts of scarce and expensive energy.

In any discussion or study of energy conservation in any manner or form, it is essential to consider the potential of optimum site selection, building siting and orientation, and site planning and design, as a primary means of conserving energy. The underlying premise for this particular study is that a maximum response to or recognition of existing site conditions, processes and factors will result in optimum and less expensive conservation measures than can be achieved by any other means.

The conceptual approach in dealing with this premise entailed a state-of-the-art study of options available for passive energy conservation in site design in each of the climatic regions based on available data, research and case studies. This study is not concerned with new original data formulation or with field testing or evaluation of existing material or research techniques. The concept of the study was predicated on the fact that much had been gathered or written in the past in relation to the climatic impact of site elements, or orientation or siting of buildings. These studies were often done in many other fields and were not directly applied to energy conservation at the time they were done or since that time. These materials have not been readily available to those now making decisions on site design or planning for energy conservation. The concept here was further expanded to assume that there were a number of studies either under way or recently completed in offices, agencies or organizations, which were not generally known to a larger audience of potential users. In addition it was felt that there would probably be actual projects either under way or completed which could be used as illustrative case studies with lessons to be conveyed to other designers, builders, developers or administrators.

The process utilized to express this concept was as follows:

1 The gathering of already existing research through a literature search and contact with associations, trade and professional associations, and with agencies, institutions and offices;

2 The location of existing studies, projects or case studies, either under way or recently completed, dealing with land planning for energy conservation;

3 The development of basic general principles for land planning or site design based either on existing research or on exemplary projects or case studies;

4 The utilization of the typical site planning process to show where in the process the optimum energy conservation decisions may be made;

5 The theoretical application of these principles and this process to each of the four major climatic regions in the continental United States;

6 The illustration of the application of the basic principles in the regions through actual case studies which were discovered during the course of research on the project;

7 The graphic depiction, where possible, of ideas, concepts or applications in order to make these comprehensible to an interdisciplinary audience more accustomed to graphic than verbal communication;

8 The provision for additional research opportunities and options through a review of existing literature, as well as the provision of a bibliography on the subject.

Thus this summary report of the state-of-the-art study is organized in two sections. The first part deals with the general guidelines for site design and energy conservation. This section deals with the relationships between the sun and wind, landforms, water and vegetation, and the resultant climate control options. The second part of this summary document deals with regional site design guidelines. In the cool and temperate regions it is often more important to conserve energy by curtailing heat loss and by utilizing passive solar energy during the extended winter season. In the hot-arid and hot-humid areas it is possible to conserve energy by utilizing natural cooling through solar radiation control and by increasing and directing the natural cooling effects of wind flow. Therefore, the site planning and design principles for each of these regions are distinctive and are treated in some detail in the text. The organization of the presentation for each of the regions is standardized, dealing the the *climatic characteristics, design criteria, gross site selection guidelines, discrete site selection guidelines, design guidelines, and use of natural and man-made elements.*

contents

1 site design & energy conservation

CLIMATE AND ITS VARYING EFFECTS

CLIMATE CONTROL OPTIONS

2 regional site design guidelines

COOL REGION

HOT HUMID REGION

TEMPERATE REGION

HOT ARID REGION

1

site
design
and
energy
conservation
climate and its varying effects

The premise of this study is that maximum response to existing natural site conditions will result in energy conservation. This premise is based on the following factors identified during data collection.

Energy Consumption For Space Heating & Cooling Is Significant

Almost 18 percent of the national consumption is used for heating residential (11 percent) and commercial (6.9 percent) buildings. The annual growth rate of fuel consumption for space heating is 4 percent (AIA 1974 p. 61).

While only 3 percent of the total national energy consumption is used for cooling, energy consumption for interior space cooling during the hot months is 42 percent (AIA 1974 p. 75).

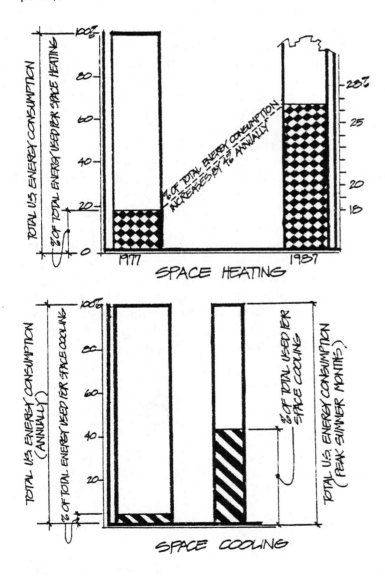

Natural Climate Characteristics Determine Heating & Cooling Requirements

Temperature, radiation and wind effects are the most significant natural conditions which have an effect on heating and cooling requirements of structures in the four major climatic regions of the United States. How warm or cold you feel in a room depends about half on the air temperature registered on your thermometer, and half on the temperature of walls, ceilings and floors not shown on your thermometer at any time of year.

The material forces that heat and cool your walls, ceilings and floors are the *sun* and *wind*. The sun's radiant heat effects are present in both summer and winter, both day and night. Sun radiation is responsible for half of the interior heat of a structure.

The wind velocity and pattern follow the same seasonal variation. The cooling effect of wind can contribute to as much as 20 percent to the interior cooling of a structure. The objective of site design for energy conservation is to respond to regional and site specific sun and wind patterns on a daily and annual basis. The goal of site design for energy conservation is to utilize or diminish these heating and cooling effects as desired.

As the above statements reflect, and as our common sense tells us, the main concern of people in the design and use of shelter is the requirement of human beings to maintain a desired level of comfort—neither too hot nor too cold.

In surveying data on climate, human comfort and design, the basic picture emerges that the primary considerations are sun and wind and the varying requirements for heating and cooling which they create. For this reason, the first part of this study will provide a cursory summary of relevant principles of climate and the manner in which existing surface conditions (landform, water, vegetation and ground surface) create variations in climate. It will also examine the various techniques by which desired effects of heating and cooling can be created.

In considering sun and radiation, it is important to understand the type, amount and degree of radiation which has an effect on the earth's surface and on man. To begin with, sun and radiation are the main source of the earth's natural heat. However, not all of the sun's radiation reaches the earth.

A considerable portion of the sun's total radiation (33 percent) is reflected by the surface of clouds and is ineffective for heating the air and the ground. In the atmosphere another 42 percent of the sun's radiation is scattered and diffused and deflected from heating the earth.

Finally, radiation is lost through absorption caused by ozone, water vapor and carbonic acid. Despite these losses, radiation does reach the earth's surface either as direct solar radiation or as scattered radiation from the sky (Geiger p. 4).

SUN AND RADIATION PRINCIPLES

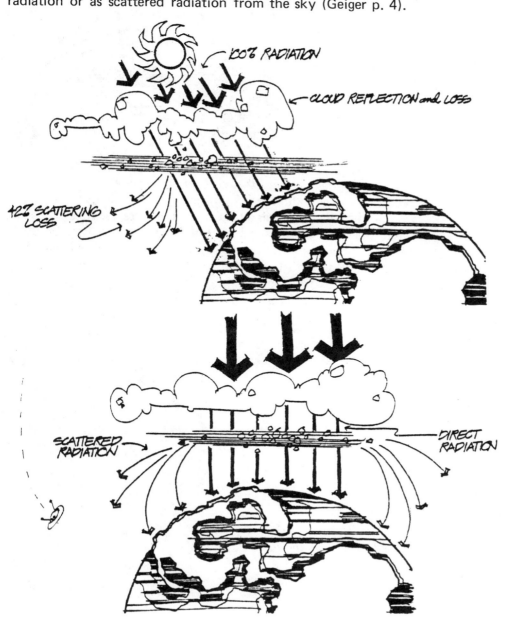

Upon striking the earth's surface, a portion of the radiation is reflected from the surface and lost for heating purposes. However, most of the radiation reaching the earth's surface is absorbed, changed to heat and raises the temperature of the ground, the air, and surrounding objects (Geiger p. 4).

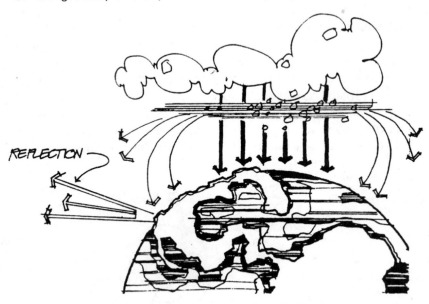

REFLECTION

Ground temperature is therefore determined by the amount of heat which the ground surface absorbs while air temperature is indirectly determined by the amount of radiation which heats the air. Because this process occurs on a daily basis, the earth's surface and the air layer near the ground become significant areas where temperature control can be affected. In terms of its effect on human comfort and design, the earth's surface (particularly during the noon day heat transfer) and its effects on the layer of air near the ground are the most significant factors (Geiger p. 4).

Classic climate studies have shown that variables affecting radiation effects on temperature are:

weather conditions: clear versus cloudy

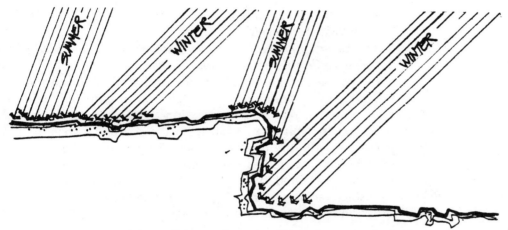

position of the surface: horizontal versus vertical

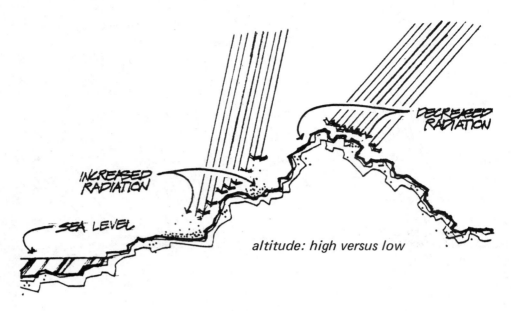

altitude: high versus low

■ *solar radiation increases with altitude above sea level in the lowest air levels*

■ *however, solar radiation decreases with altitude above sea-level in the higher air level (Geiger p. 4).*

So far we have looked at radiation patterns during the day, however, they should also be considered in terms of their nighttime patterns. While solar radiation during the day is dependent upon incoming radiation from the sun and associated heat exchange with the ground surface and air, the process at night can be summarized as the reverse of the daily phenomenon. Most simply stated, heat exchange during the day is due to incoming solar radiation, while heat exchange at night is due to heat radiation from the surface of the earth (Geiger p. 14).

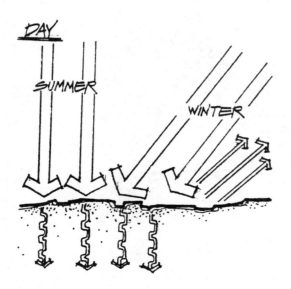

Just as with daytime radiation, certain variables affect nighttime radiation and can be summarized as follows:

Radiation Loss: The degree of nighttime radiation loss is affected by the atmosphere's water content. Only 12 percent of the evening radiation is lost to free space. The remaining radiation is absorbed by the various layers of air in a degree proportionate to their water vapor and carbon dioxide content (Geiger p. 14).

Cooling: As hot air rises, the ground and air surface close to it begin to cool and cool the air mass above. This cooling extends vertically several hundred meters (Geiger p. 23).

 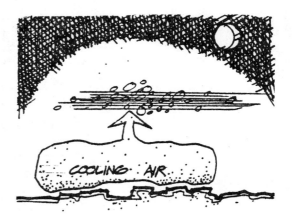

The heat loss or cooling effect is greatest for the air near the ground and decreases with distance from the ground as the cold heavy lower air layers become trapped by warm lighter upper air layers. The significance of the daily pattern of incoming (daytime) and outgoing (nighttime) radiation is due to its effect on daily and annual temperature patterns. Measurements on the effect of radiation patterns on temperature and shelter have been carried out by various classic climatic studies and can be summarized as follows (Geiger p. 63).

- increase of daily temperature range with approach to the ground is common to all macro-climates of the earth (Geiger p. 77).
- cloudiness influences daily temperature according to the extremes of summer (June) and winter (December) as follows.

In Summer: Cloudy weather only slightly reduces temperature.

In Winter: Cloudy weather causes a rise of temperature while clear weather brings frost (Geiger pp. 78-79).

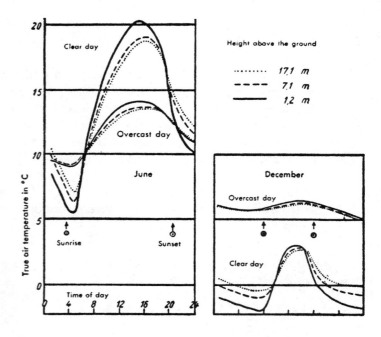

The varying changes of temperature and the effect of cloudiness on temperature are due to basic laws of physics. They explain heat transfer and radiation which affects climate in basically the same way in which climate affects the heating and cooling of structures.

Radiation: Every body emits radiant heat in accordance with its own temperature. Radiation during the day is a process of incoming radiation flow which causes heat gain. Radiation during the night is a process of outgoing radiation flow which causes heat loss.

The basic principles involved are based on molecular flow.

Conduction: Warm bodies give warmth to cold bodies and lose their own heat.

Convection: Masses of liquid and gas effect temperature change by displacement of the heat energy of their surroundings which they carry with them (Geiger p. 26).

Essentially, this is how exterior temperature causes structures to be heated and cooled. In the summer and winter, if radiation is permitted to enter a structure, it causes interior temperature to be raised.

In order to understand the varying requirements of heating and cooling created by the sun and radiation, it is important to understand the manner in which radiation varies according to seasonal and daily requirements and existing natural conditions. These variations are summarized in the following section.

Though the specific characteristics and effects of radiation vary with different geographic regions and climatic zones of the United States, the following are variations which hold true for every climate region of the United States.

RADIATION AND ITS VARYING EFFECTS

Seasonal Sun Angle: The noon summer sun angle is 45 degrees higher in the sky than the noon winter sun.

The resultant effect on structures is that winter noon sun comes further into the interior of a structure, while the noon summer sun is easiest to stop. (Wright Oct. 49 pp. 158-159).

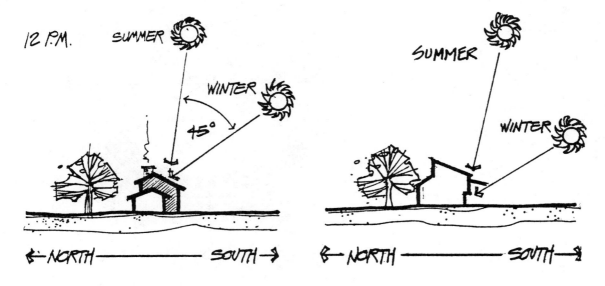

Seasonal Path: In the temperate zone, the daily degree of the sun varies as follows: In summer, the sun path from sunrise to sunset is a full 240 degrees so that east to west facing surfaces receive more sun than the south or north. The horizontal roof surface receives a greater amount of radiation than all other sides of a structure. (Olgyay p. 87).

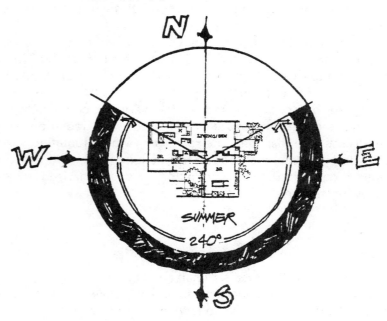

In winter, the sun's path from sunrise to sunset is 120 degrees so that the greatest amount of sun is received on south surfaces (Wright p. 159).

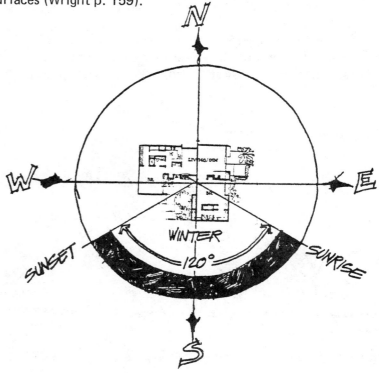

Site related implications of these variations are:

- slopes with southern orientations are always considered the optimum slopes for residential developments.
- site design for reduced winter heating requirements should optimize the winter south sun and minimize prevailing wind through siting, orientation and design of structures, activity areas, site circulation and site details.
- site design for optimum summer cooling effects should minimize the eastern and western sun and optimize cool breezes through siting, orientation and site details.
- for a year around balance, use southern sun and avoid western and southwestern sun (Wright p. 158).

Seasonal Sunrise: The other variable is the daily pattern of sunrise and sunset which changes with the season. In winter, sunrise is to the east, sunset is to the west. In summer, sunrise is to the northeast and sunset to the northwest.

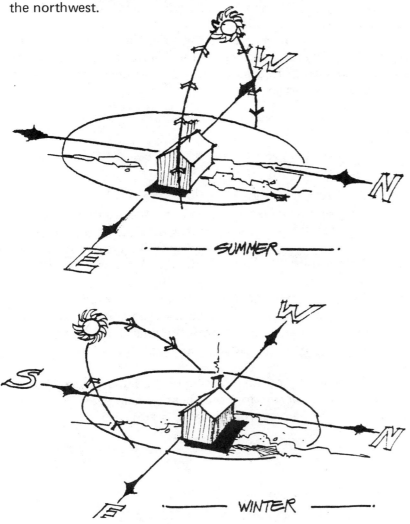

SUMMER

WINTER

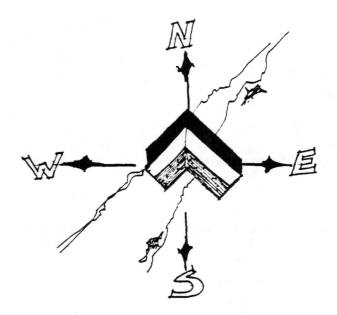

The significance of this data is applicable to the configuration and internal layout of structures. Suggested form and orientation is an "L"-shaped plan with major windows facing southeast (Wright p. 159).

Essentially these are the characteristics of the sun which combine with natural factors of slope, water bodies, vegetation and ground surface to suggest desirable site characteristics with implications for design.

RADIATION AND VARYING SURFACE CONDITIONS

Topography And Landform

During the day the topography has a great effect on climate in that the sun delivers different quantities of heat to sloping and flat ground. To what degree the ground is favored or on the contrary, depends on the directions and inclination of the slope (Geiger p. 125).

While regional and site specific determinations frequently can only be made on the basis of sun angle calculations, radiation charts and climate data for annual diurnal variations, the following generalities about topography and radiation can be made.

Slope aspect or the direction of the slope in relationship to the sun is important for solar heat gain in every region of the country. On a year-round basis, south facing slopes are always the hottest.

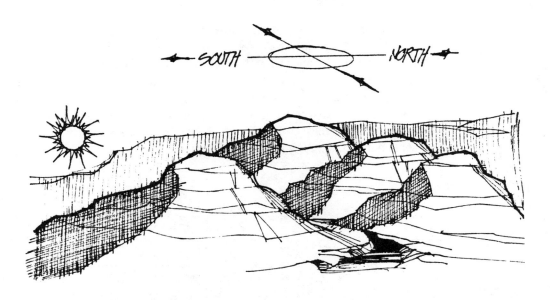

General slope characteristics for every region of the United States are as follows:

Southeast Slope: Most desirable.
South Slope: Preferred.
■ warm winter
■ early spring
■ late fall
East Slope: Acceptable
■ warm winter mornings
■ cool summer evenings
West Slope: Undesirable
■ hottest summer slope
North Slope: Least desirable
■ coldest in winter

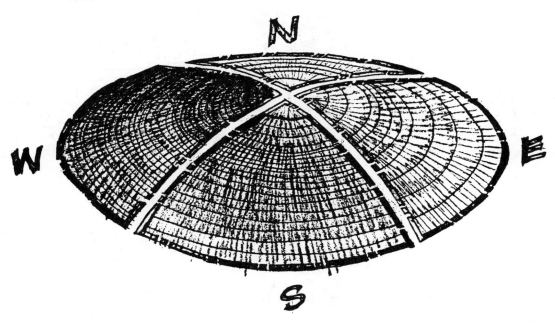

Measurement of Climate Modification

Studies in the New York—New Jersey region have measured the impact of radiation on slopes of varied orientation and degree of steepness in order to evaluate total radiation effect for site selection purposes.

Data surveyed indicates that preferred slope direction and gradient can produce the following extension of desired seasonal effect (Olgyay pp. 49-50).

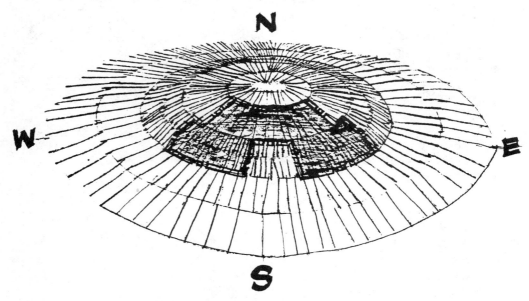

Slopes of 10 percent gradient with south and southeast orientation receive 20 percent or more solar radiation and will be two weeks ahead of any other slope gradient or direction in the arrival of spring.

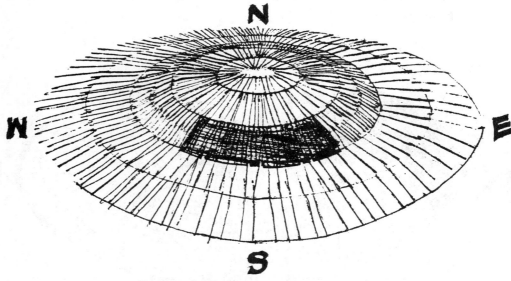

Slopes of 20 percent gradient with southern orientation receive 30 percent more solar radiation and will be three weeks ahead of any other slope gradient or direction in the arrival of spring.

Slope Direction And Angle

The amount of radiation received by topography depends on the slope of the land. Maximum direct radiation is received by a surface that is perpendicular to the direction of the sun. A site that faces the sun squarely receives more sun rays than one edgewise to the sun. For example, a 17½% slope at 40° N latitude on a cloudless day attains the following approximate percentage of radiation (Lynch p. 68).

Orientation	Midsummer	Equinox	Midwinter
North	95%	55%	15%
East or West	100%	60%	25%
South	100%	70%	35%

Difference in slope direction and angle can affect site micro-climate in the same way that differences in latitude affect the climate regions of the United States overall. Measures of effect on climate modification due to slope direction and angle are as follows:

Ohio: A south facing slope of 1:12 versus a north facing slope of 1:12 gradient can produce differences in site micro-climate as illustrated below (Langeweische Oct. 1949 p. 147).

The implications of this data are that slope direction and gradient do effect site micro-climate significantly.

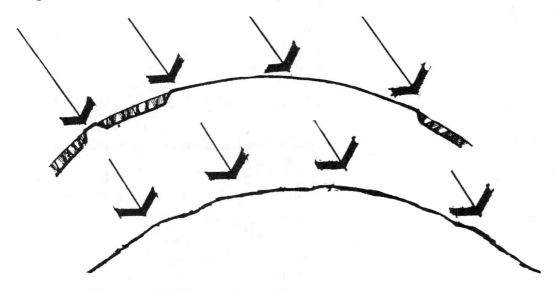

Regional Variations

Dependent upon variation in regional climate and regional climate requirements, the following variations in radiation impact on vertical wall surfaces should be considered in siting, planing, and buffering structures.

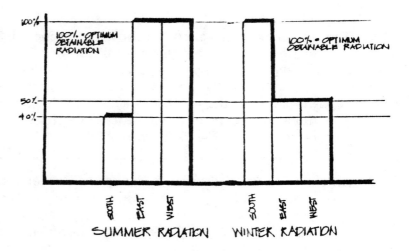

SUMMER RADIATION   WINTER RADIATION

Cool and Temperate Zones: In winter, south facing walls receive twice as much solar radiation as east or west walls. In summer, east and west facing walls receive 2½ times the solar radiation of south facing walls (Olgyay pp. 86-87).

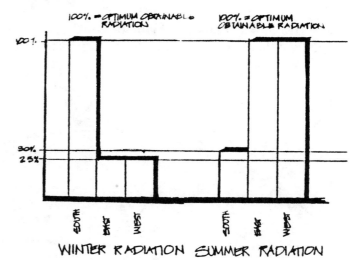

WINTER RADIATION   SUMMER RADIATION

Hot Arid and Humid Zone: In winter, south facing walls receive 4 times as much solar radiation as east or west walls. In summer, east and west facing walls receive 2 or 3 times as much radiation as south facing walls. In the lower latitudes of this region, summer radiation impact is twice as great on north versus south walls (Olgyay pp. 86-87).

Location Above Or Below Ground

As the discussion on radiation and ground surface indicates, radiation on the earth's surface is either absorbed into the ground or reflected into the lower air levels. An area of consideration is the correlation between temperature differences above and below the ground.

16

Classic studies on temperature and climate have shown that above ground temperatures are consistently hotter in summer and cooler in winter, while the reverse situation is true below ground (Langeweische Aug. 1950 p. 92).

Weighing this data against other regional climate, a comparative analysis indicates the following regional suitability of completely or partially subterranean construction (Langeweische Aug. 1950 p. 93).

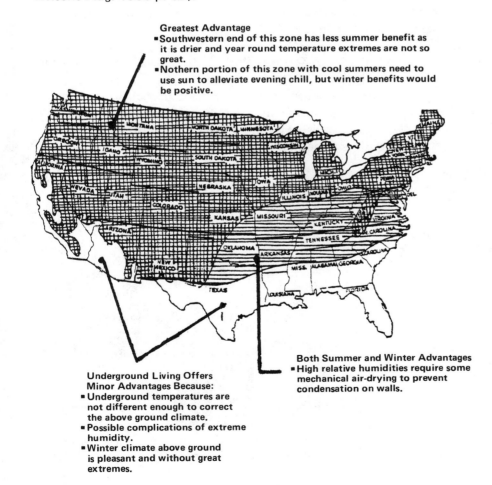

Greatest Advantage
- Southwestern end of this zone has less summer benefit as it is drier and year round temperature extremes are not so great.
- Nothern portion of this zone with cool summers need to use sun to alleviate evening chill, but winter benefits would be positive.

Underground Living Offers Minor Advantages Because:
- Underground temperatures are not different enough to correct the above ground climate.
- Possible complications of extreme humidity.
- Winter climate above ground is pleasant and without great extremes.

Both Summer and Winter Advantages
- High relative humidities require some mechanical air-drying to prevent condensation on walls.

Surface Reflectivity

Another factor affecting radiation is the type of surface which the radiation strikes. Essentially, this characteristic involves two principles. One deals with the characteristics of sunshine, the other with the reflective characteristics of various materials. They can be summarized as follows.

Sunshine: Essentially the sunrays which affect the earth and its climate are of two types—ultra violet rays and infra red rays. Both of these types of rays are reflected by whatever surface they strike, however, infra red rays are heat rays, and have a greater impact on the surface they strike and should be considered more carefully (Wright Oct. 1949 p. 153).

Albedo: This term refers to the characteristic of any surface which reflects back, rather than absorbs radiation. Both natural surfaces, such as snow, water, earth, etc., and man-made surfaces such as asphalt, metal, etc., and surface colors possess varying albedo characteristics.

The interrelationship between radiation, albedo, and climate impact can be simply illustrated in the following example of the effect of varying surface paving exposed to western summer sun.

Dark Paved Terrace: With a dark paved terrace on the west side radiation is absorbed by the paving while infra red (heat) rays are reflected against the house. This reflection heats the ground surface and the house's surface walls, windows and roof simultaneously. The result is undesired summer heat gain.

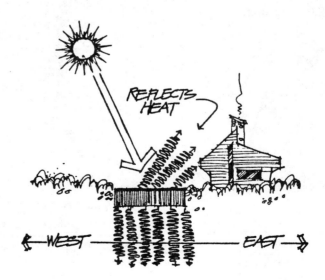

Light Paved Terrace: With a light colored paved terrace, radiation is absorbed by the paving at a slower rate while glare is undesirable for its effect on vision, heating impact is not so great.

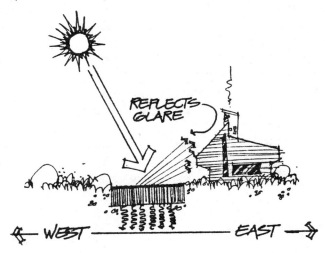

Essentially neither situation is desirable. Instead, selection of a natural or man-made material which is exposed to intense sun should be based on the albedo of the surface. Extensive studies of albedo have identified the following characteristics of surface materials which should be considered.

REACTION OF MATERIALS TO SOLAR AND THERMAL RADIATION

	Percent of Reflectivity		Percent of Emissivity
	Solar Radiation	Thermal Radiation	Thermal Radiation
Silver, polished	93	98	2
Aluminum, polished	85	92	8
Whitewash	80	—	—
Copper, polished	75	85	15
Chromium plate	72	80	20
White lead paint	71	11	89
White marble	54	5	95
Light green paint	50	5	5
Aluminum paint	45	45	55
Indiana limestone	43	5	95
Wood, pine	40	5	95
Asbestos cement, aged 1 year	29	5	95
Red clay brick	29-30	6	94
Gray paint	25	5	95
Galvanized iron, aged (oxidized)	10	72	28
Black matte	3	5	95

Albedo for Visible Portion of Spectrum

Fresh snow cover	80–85%
Older snow cover (wet)	42–70%
Fields, meadows, tilled soil	15–30%
Heath and sand	10–25%
Forests	5–18%
Surface of sea	8–10%

(Geiger p. 129)

(Olgyay p. 114)

Other characteristics of surface materials related to reflection which should be considered are as follows:

Moisture: The albedo of moist surfaces is less than that of dry surfaces (e.g. dune sand has an albedo of 37 percent when dry, but only 24 percent when wet). (Geiger p. 129).

Position: Roughly twice as much radiation falls on a horizontal surface during overheated periods than on a vertical surface. Horizontal surfaces reflect a large portion of heat into surrounding buildings during overheated periods.

Surface Temperature Variation: Exposed to the same sun, varying surface temperatures have been shown in the following range (Langeweische Oct. 1949 p. 192).

Material	Temperature	% Increase Over Minimum
Trees	80°	Minimum
Lawn	110°	38%
Wall Surface	130°	63%
Asphalt	160°	100%

Grass Surface: 33 percent cooler than paving exposed to the same sun. Grass always keeps a constant temperature because the sun's heat disappears as fast as it hits green grass. You feel cool because you are radiating heat to the grass (Langeweische March 1950 p. 108).

White Roof: 10° − 20° cooler than dark roof exposed to the same sun (Langeweische Oct. 1949 p. 152).
Regional Climate Variation: Temperate difference between city and surrounding suburbs has been measured to be almost 13 percent less (Langeweische October 1949 p. 147). Differences in temperature are due to the low albedo of suburban lawn and tree cover, as well as the greater absorption of solar radiation by urban pavement and buildings. Added consideration should be given to the temperature differences created by the urban heat island phenomenon.

Water Bodies: Surface Absorption And Radiation

Water bodies store more insolated solar energy and radiate less energy than surrounding land masses. Proximity to water bodies has been found to moderate temperature extremes of adjacent land forms by raising winter temperature and lowering summer temperature (Olgyay p. 51, ASLA p. 61). Water temperature variations have been identified in a range of 20° − 30° difference for summer and winter (ASLA p. 184).

Great Lakes Region
Winter heat gain − 5° F average
Summer heat loss − 3° F average (Olgyay p. 510)

Vegetation

Vegetation affects radiation by blocking direct solar radiation and reducing heat loads on exposed surfaces. Vegetation shades because plants absorb rather than reflect sunshine. Vegetation can also improve summer micro-climate by filtering and cooling the air (Davis and Schubert p. 43).

The measurement of vegetation as a tool for climate modification has been surveyed and is summarized as follows:

Degree of Sun Absorption: Shading from trees can prevent as much as 70 percent of the sun's heat from being absorbed by the ground, and this combined with the trees' transpiration process will have a measurable effect on the air temperature around them (Kerner p. 54).

Effect on Air Temperature: Air temperature can be cooled 10 to 20 degrees due to shading when the general temperature is 90 degrees. Air temperature can be cooled by 5 to 10 degrees when the general temperature is 70 degrees.

Effect on Ground Temperature: In a forest situation 80 percent of the incident solar radiation is caught in leaves, needles, twigs and branches of vegetation, so that less than 5 percent reaches the forest floor during the day (Geiger p. 317).

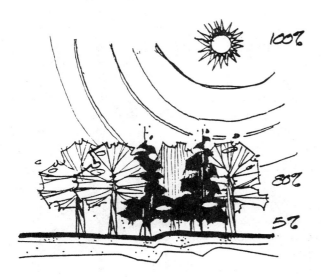

Cooling Effect: During the day, the cooling effect of vegetation has been demonstrated as follows. At sunrise, it is coolest in the tree crown with a cool air layer maintained on the ground surface under it. For three hours, until the sun reaches noon level, the ground remains cooler and is not heated until radiation penetrates the entire canopy.

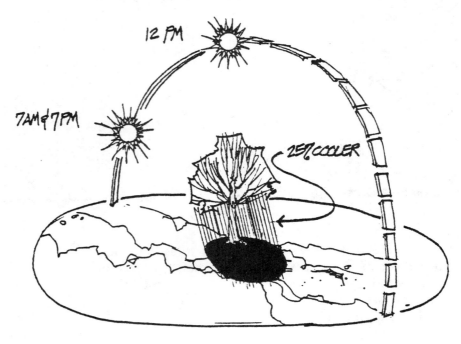

Observed temperature differences have shown the ground surface to be at least 25 percent cooler than the air above the crown (Geiger p. 333).

During the evening the outer surface of vegetation is the primary cooling agent and throws off the greatest amount of radiation. The outer surface is 2.5 degrees cooler than the surrounding air temperature and is almost 25 percent cooler than the lower part of canopy (Geiger p. 325).

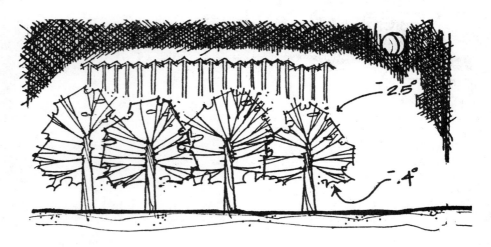

Other measured effects of vegetation on radiation and climate modification have been observed as follows.

Immediate surroundings of a house make a big comfort difference. Trees here kept ground from heating up. As sun sets, the house cools off soon, and lets you sleep. Bare pavements have heated up during the day. After sunset, hot pavements keep radiating heat at the house, keeping lower floors hot until after midnight (Langeweische Oct. 1949 p. 148).

Type of Vegetation: The type of vegetation used will affect the type and degree of radiation control.
- Evergreen—reduces light penetration to 8 percent.
- Deciduous—reduces light penetration in range from 51 percent to 4 percent (Hastings and Crenshaw p. 1-9).

Shrubs and ground cover reduce temperature due to absorption and evaporation. Measured summer cooling effect has shown 10 to 14 degrees cooler temperatures for grass surfaces as opposed to exposed soil (Olgyay p. 51).

WIND AIR FLOW PRINCIPLES

Wind moves large masses of air across the surface of the earth. This movement occurs in regular daily and seasonal cycles. The significant aspect of wind is that it increases evaporative and convective cooling. These principles are probably most easily explained in terms of their individual effects on climate and their joint effect on heating and cooling requirements.

Convection: Affects temperature by the displacement of heat by means of liquid and gases which carry surrounding heat energy with them.

Evaporation: Affects temperature by changing the moisture content of air and ground and carrying surrounding heat energy with it (Geiger p. 26).

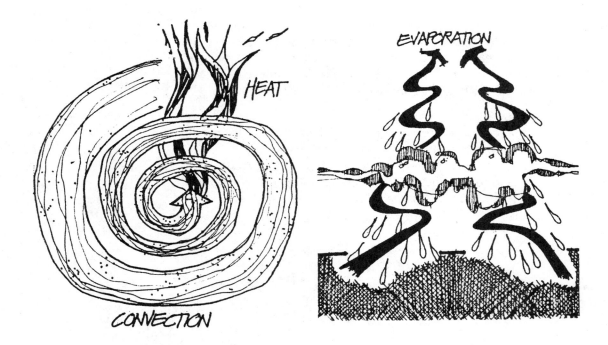

Convection and evaporation affect heating and cooling requirements by changing the temperature of air and the drying capacity of breezes.

In every part of the country, air flow and wind velocity vary according to the following conditions.

Daytime Airflow: During the day, air directly at the ground surface is at rest. Through the breaking and eddying of wind on the ground, ground air moves up carrying with it a lesser horizontal motion. When the slower air levels meet faster upper air levels, a breaking effect occurs. The higher more mobile air then flows into the lower level.

Nighttime Airflow: At nighttime when upper air is at rest, the force of the air movement generally diminishes. Since there is less eddying diffusion, nighttime air is generally calmer than daytime air (Geiger p. 102-115).

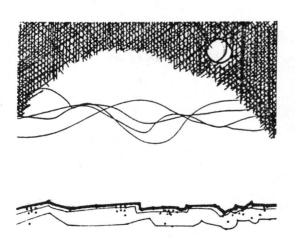

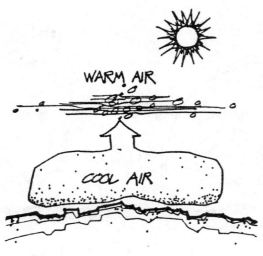

Air Temperature: Air of low temperature is heavier than air of higher temperature. Cold air tends to push itself under warm air and creates air circulation until equilibrium is achieved (Geiger p. 195).

Cold Air Floods: These are a result of air temperature differences. For example, at night in hilly country cold air from high places flows to low places and is replaced by warm air from low places.

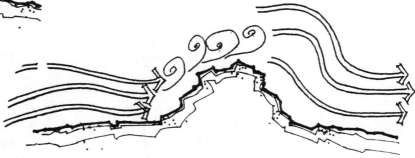

Cold Air Flows in Valleys: Cold air movement gains significance when it occurs in great volume (Geiger p. 204). In large valleys, cold air flow at night is due to outgoing radiation from the valley floor and from the side slopes (Geiger p. 211).

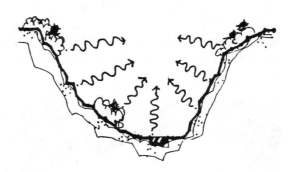

The most violent wind of the free atmosphere is to some extent slowed down by the ground. Directly at the surface, the air is entirely, or almost at rest. The movement of wind is effected in the following ways.

Air Flow: The breaking effect of the ground causes eddy diffusion and transmits air up. The nearer the air to the ground, the more all air movement is hindered.

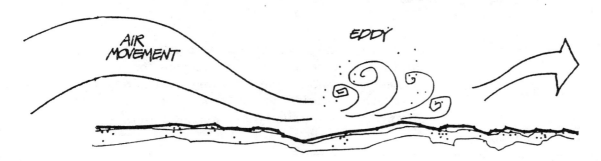

Wind Velocity: The air near the ground is the part of the atmosphere where wind velocity increases with height (Geiger p. 102). The primary factor influencing the variation of wind velocity with height is temperature. When temperature decreases with height, wind velocity increase is less than when temperature increases with height (Geiger p. 106). The influence of wind velocity on temperature is due to the fact that high wind velocity increases convection. This results in a decrease of ground temperature by day and increase of ground temperature by night.

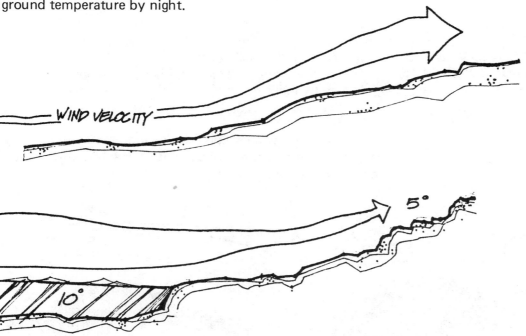

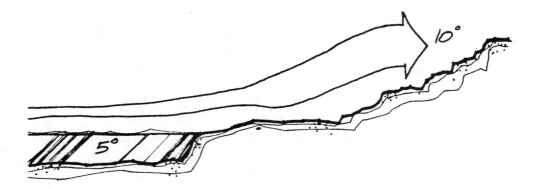

The main significance of wind velocity is that it can be correlated with temperature requirements for human comfort. Studies have shown the following correlation between exterior wind velocity and interior temperature required to offset cold air infiltration (Langeweische p. 88).

Wind Speed (mph)	Heating Requirements
1/10	68°F
1/2	73°F
1	75°F
2	77.7°F
3	78.3°F
5	79.3°F
10	80.6°F

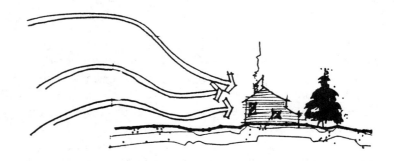

WIND AND VARYING SURFACE CONDITIONS

Hills and Wind Velocity: Hills affect horizontal and vertical windflow causing higher wind speeds at the top and windward sides of a hill and less turbulence on the lee side.

Greatest wind velocity is in the area at the sides and below the crest. Lowest velocity is at the bottom of the hill and in the wind shadow (Olgyay p. 50).

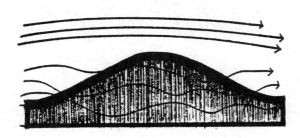

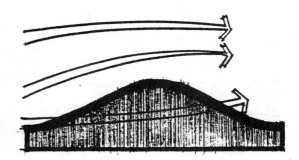

Valleys and Cold Air Pockets: Cool air is heavier than warm air and forms a cold air layer near the ground at night. In valleys, surrounding slopes cause warm and cool air to mix and create a warm slope region.

 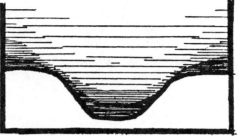

As a result, the floor of the valley and upper plateau are the coldest locations, while the warm slope region, unless exposed to the wind is the warmest location (Olgyay p. 45).

1. *Toronto, Canada*
   Valley bottom—34° colder due to cold air lake effect.

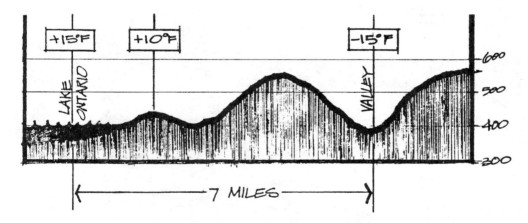

2. Frost holes
   11° colder than surrounding terrain (Davis and Schubert p. 29).
3. Wind speed on the crest of a ridge is up to 20% greater than wind speed on flat ground (Lynch p. 74).

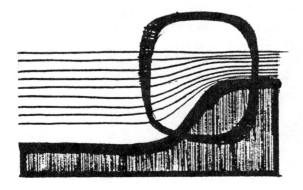

Water Bodies

Water bodies also affect the temperature of adjacent land masses on a daily basis due to the process of air flow. During the afternoon, when land is warmer than water, low cool air flows over the land and cools it. During the evening, the ground surface begins to radiate heat and cool off. At the same time, warm air above the water rises due to convection and is replaced by cool air flowing in a low pattern along the land's surface.

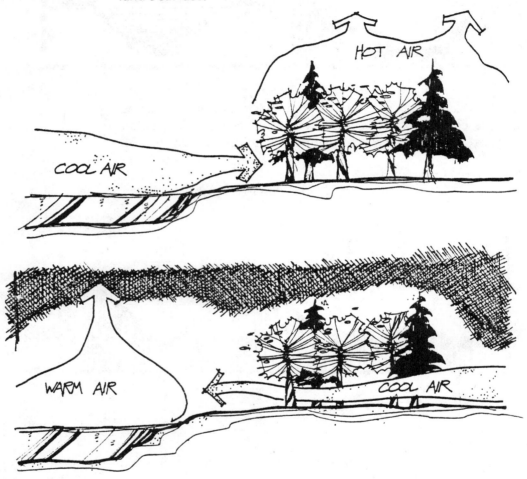

The degree to which this air flow is effective is dependent upon the size of the water body, and location on the lee side. In general, the effect of cooling is proportionate to the size of the water body (Olgyay p. 51).

1. *Great Lakes Region*
   Daytime cooling effect of water 10° F. average. (Olgyay p. 51).

Vegetation

Though large air masses cannot be altered in motion, they can be reduced in velocity near the ground by the presence of vegetation. Essentially, vegetation reduces wind velocity due to frictional drag. Vegetation is an effective wind break because it permits jet air movement through it (Olgyay p. 78).

Effect on Velocity: Vegetation provides a more extended area of protection than any other shape. According to classic climate studies (C. G. Bates) while various forms provided greater percentage of wind reduction over a limited distance, vegetation provided the greatest extended wind reduction when compared to other forms which were impenetrable (Olgyay p. 68).

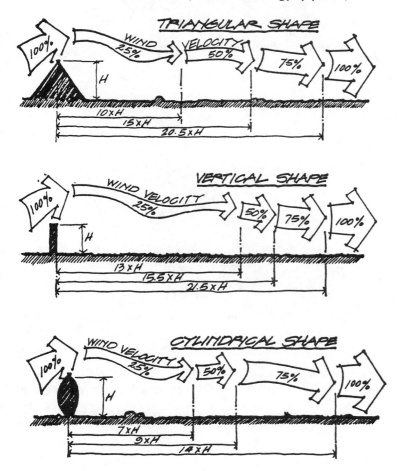

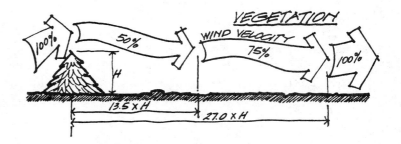

For the most extended area of wind protection, trees are the most effective windbreaks.

Influence on Air Flow: Studies on vegetative windbreaks indicate that vegetation affects air flow in the following ways.

The first effect is that vegetation diverts air current upward on the windward side.

While the windflow does turn back and sweeps the ground, an area of calm is created near the ground on the lee side.

The most protected part of the "sheltered area" is close to the windbreak on the leeward side with a small protected zone on the windward side—especially if the windbreak is dense.

If it is so open that the wind can sweep under the trees, the windward side has little protection.

Vegetation can reduce wind velocity by 50 percent over a distance from 10 to 20 times the height of the windbreak. Vegetation can affect wind velocity on both sides of the wind break. For example: A 30' high windbreak perpendicular to the prevailing wind will reduce wind velocity for 100 yards on the windward side and 300 yards on the lee side (ASLA pp. 72-78).

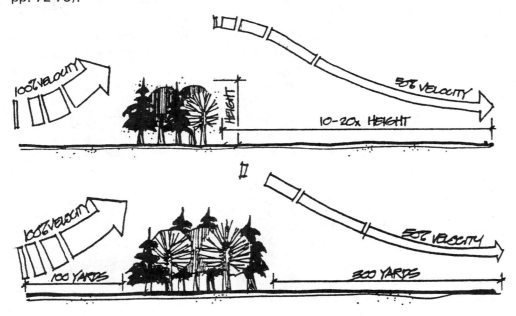

Measure of Climate Modification for Windbreaks

1. *Kansas*

   Air temperature reduction was recorded between 2 and 5 degrees cooler on the lee side of a windbreak and 6 degrees warmer on the exposed side for a distance of 6 to 24 times the height of the windbreak (Read p. 4).

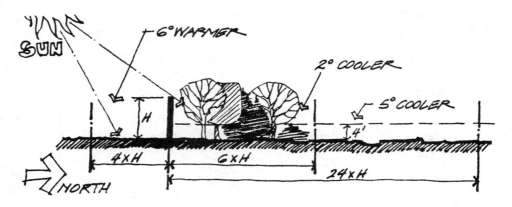

2. *Great Plains*

   Increased humidity

3. Reduced evaporation on lee side of windbreaks can extend to a distance 24 times the height of the windbreak on the lee side and reduce water loss from reservoirs and ponds (Read pp. 4,5).

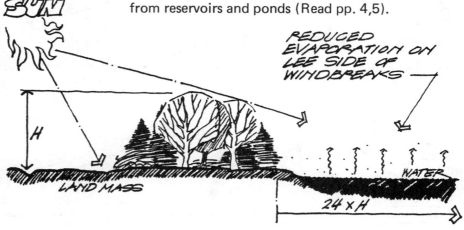

climate control options

The sun can contribute up to 50 percent of the interior heat of a structure. The sun's heating effect on a structure is due to the amount of radiation absorbed in a structure's roof, walls and ceilings which causes a rise in interior temperature. By proper planning and design, you can use this natural heat to heat you in the winter and still keep it from heating you excessively in the summer.

Maximum response to sun and wind through orientation and seasonal sun shading and wind buffering has been modeled and demonstrates the following energy savings (Olgyay pp. 132-151).

RADIATION CONTROL

SOUTH — HIGH SUMMER SUN

SOUTH — LOW WINTER SUN

MEASUREMENT OF ENERGY SAVINGS

Winter	Cool	42% reduction of heat loss
	Temperate	49% reduction of heat loss
	Hot Arid	39% reduction of heat loss
Summer	Cool	75% reduction of heat gain
	Temperate	71% reduction of heat gain
	Hot Arid	42% reduction of heat gain
	Hot Humid	55% reduction of heat gain

Passive Solar Housing

Direct (passive) solar heating can be achieved through a combination of optimum southern orientation and use of double glazed south facing walls to provide interior heat through diffuse radiation (JAE). This option is immediately effective and only requires proper siting. Design guidelines for passive solar heating can be summarized as follows (AIA 1976 pp. 64-73).

Gross Site Selection Criteria: In assessing and evaluating sites, look for:

■ seasonal and daily path of the sun across the site.
■ seasonal and daily windflow patterns around and through the site.
■ avoid earthforms that block the sun or wind.
■ avoid presence of low areas where cold air can settle.

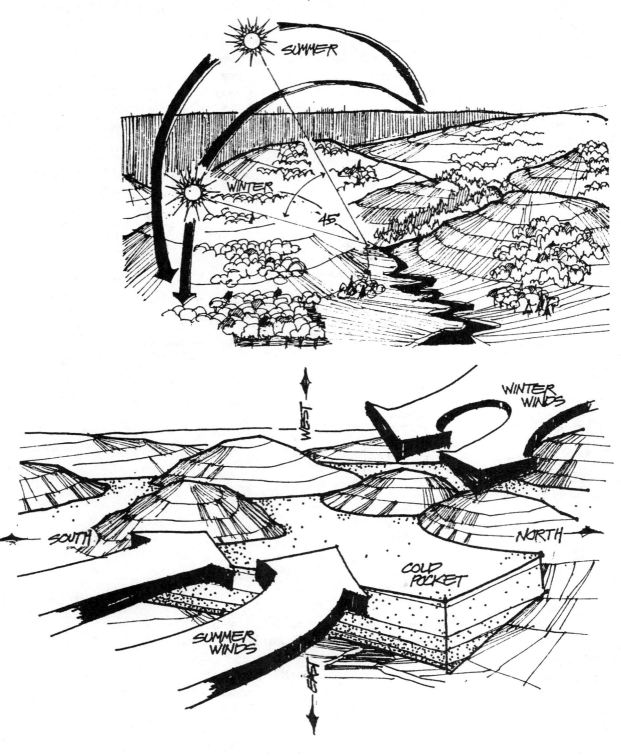

Discrete Site Selection Criteria: Look for:
- south facing slopes for maximum solar exposure
- west facing slopes for maximum afternoon solar exposure
- east facing slopes for maximum morning solar exposure
- north facing slopes for minimum solar exposure
- depth and type of rock on site
- unbuildable areas on site
- existing vegetation: type and location
- climatically exposed areas

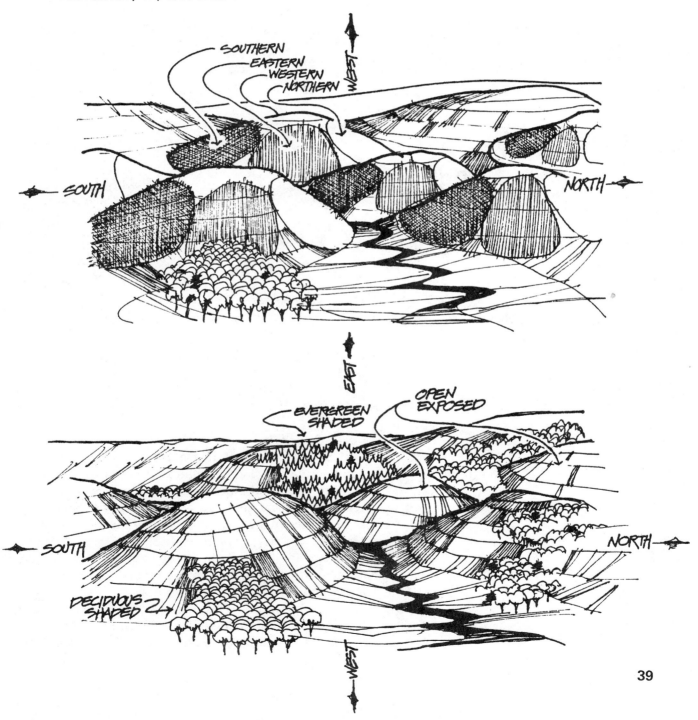

Orientation: Depending on your climate region, orient your structure for optimum summer sun.

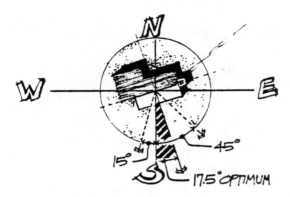

Temperate Zone: S 17° 5′ SE

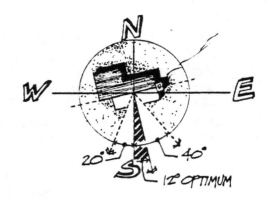

Cool Zone:
S 12° Optimum on slopes of 0-20%

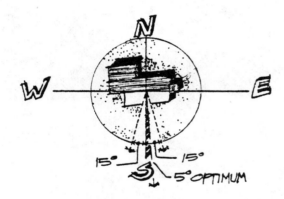

Hot Humid Zone: S 5° SE

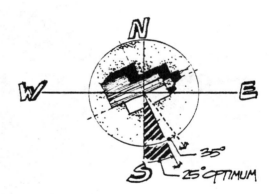

Hot Arid Zone: 25° SE

Measure of Energy Savings for Optimum Orientation

Davis, California

A 1,500 square foot house with southern exposure, good insulation and ventilation, single pane windows, a white roof, and a bare concrete slab floor was monitored on typical clear December days with an average outdoor temperature of 45° F. During the middle six hours of the day, radiant heat gain through south windows was found to eliminate the need for other heating. Total heating requirements are reduced by one fourth and produce a 54% fuel savings over a house without southern exposure. This house is 61% solar heated, requiring 39% as much energy for heating for the entire winter as the average Davis house (Hammond 1974 p. 41).

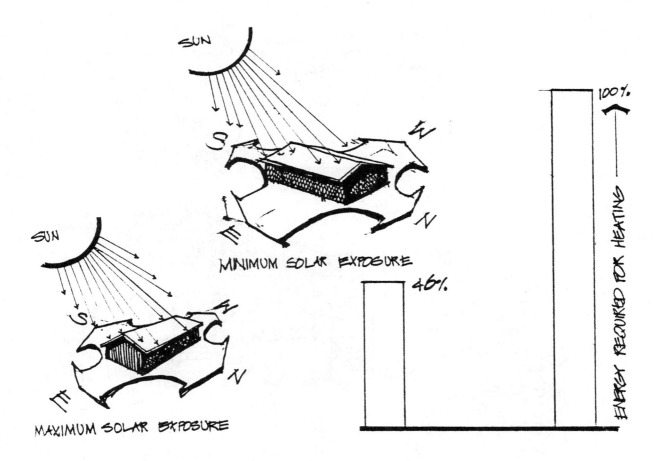

MINIMUM SOLAR EXPOSURE

MAXIMUM SOLAR EXPOSURE

A study of south facing unoccupied apartments during sunny, clear, cold days in December and January and February, compared to similar apartments with north, east, or west orientations, identified the following energy efficiency for southern orientations:

1. Interior temperatures for south-facing apartments were 24° above the ambient temperature and 17° above the north, east, or west oriented apartments.
2. Average gas use for winter months in south facing apartments was approximately 12% lower than east and west facing apartments and approximately 25% less than north facing apartments (Hammond & Hunt 1974 p. 14).

NAHB Test House

N-S orientation provided heat gain over assumed E-W orientation (NAHB p. 15).

Configuration: If you want to increase the amount of sun you get and extend your outdoor living season, consider a configuration that will create a sun pocket. Use an "L" shaped floor plan with open part facing south to create the sun pocket.

SUN POCKET 70°
EXTERIOR TEMPERATURE 30°

In one New York Metropolitan area an "L" shaped floor plan was used on a NW hill to buffer wind and create a sun pocket (Langeweische p. 150).

■ on an April day, measured air temperature in the sun pocket was 50° higher than ambient air temperature.

■ the use of dark colored asphalt paving on the patio surface is the sun pocket and extensive glazing on the surrounding walls should be considered as contributing to the temperature gain in the sun pocket.

Solar Heating Systems

Another option you might wish to consider is a solar housing system. While this entails a large initial cost for equipment, operating costs are negligible. This option is suitable for new construction or as an adaptation on existing structures. General considerations for solar houses are as follows.

Availability of Solar Energy: The basic climate variables which determine the availability of solar energy depends on:
- latitude
- season
- weather patterns

These factors are greatly influenced by the presence of fog, cloud cover, storms or smog, which can appreciably reduce incoming solar radiation.

Two basic methods to estimate actual availability of solar energy are:
- national weather service has some stations which record these types of measurement, also local universities or pollution control districts may (Leckie p. 87).
- by knowing the percentage of cloud-free days each month, you can adjust the cloud-less insolation data (ASHRAE values), accordingly.

Design Guidelines: The guidelines for site selection and design for solar housing are summarized as follows.
- minimize wind and shading during the heating season.
- utilize natural windbreaks and deciduous trees to reduce infiltration losses. Don't place large glass areas or poorly fitted doors on the windward side of the building.
- try to maximize insolation available by siting and orientation.

System Requirements: Dependent upon the type of solar collecting system you use, the following siting considerations should be considered.

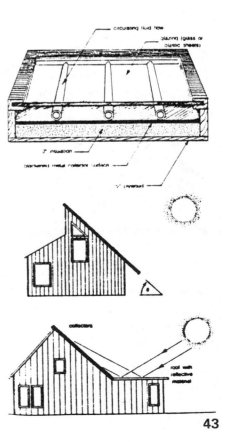

Flat Plate Collector: Siting and orientation (Leckie pp. 111-112).
- should be exposed to direct sunshine.
- operates well with diffused/indirect radiation or under hazy conditions.
- panel should always be perpendicular to incoming sunlight, (this isn't totally possible if you have a stationary collector).
- orient for sun angle during winter months, low in the southern sky (S-SW), optimum winter tilt equals local latitude plus 15 degrees measured from the horizontal.
- this requires a relatively steep roof if roof mount is desired, but the angle need not be perfect, particularly with additional reflection from another roof surface (Leckie p. 118).

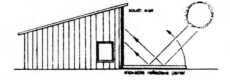

■ reflective panels may also be used to supplement.

Focusing Collectors: Siting requirements.
■ unclouded, direct sunlight is required. If it is not available, the focuser collects little useable energy. Consequently their use is limited to areas with very few days of cloud cover during the heating season (Leckie pp. 122-123).

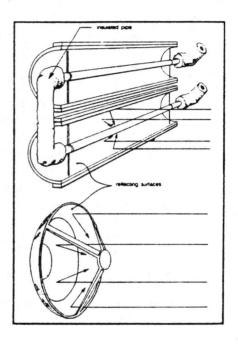

RADIATION CONTROL FOR SHADING AND COOLING

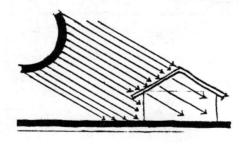

The undesireable effect of radiation is primarily the summer effect of excessive heat. Negative summer effects of radiation are due to the heating of building roofs, walls, and ceilings, which raise the interior temperature of a structure.

It is not the exterior temperature which makes your house hot, but the hot air admitted into your house and trapped there. Even if you've selected the optimum site or orientation, or if your present one is less than optimum, you will probably be interested in one or several shading and cooling options.

The main considerations in selecting an option for shading and cooling are:
■ cost versus effectiveness
■ time versus effectiveness

Shading

The primary determinant in selecting a shading device is the sun exposure to which you are responding and requirements of summer sun response versus winter sun response. Depending on your sun exposure, here are your options (Wright Oct. 1949 pp. 220-222, Siple Nov. 1949 p. 199).

Large Scale Controls: Primarily involve response to western sun to optimize winter sun and shade summer sun. These should be considered in siting, orientation and design of your house, garage and sun shields.

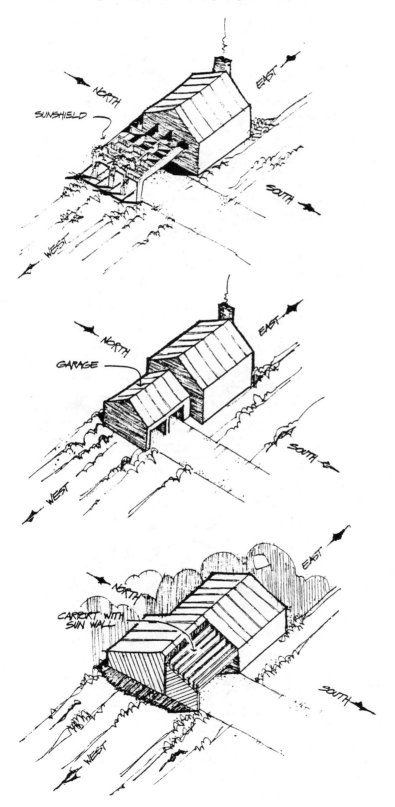

Small Scale Controls: Primarily involve response to added shading requirements for southern and western sun. Such options should be considered as the design and placement of trees and overhangs.

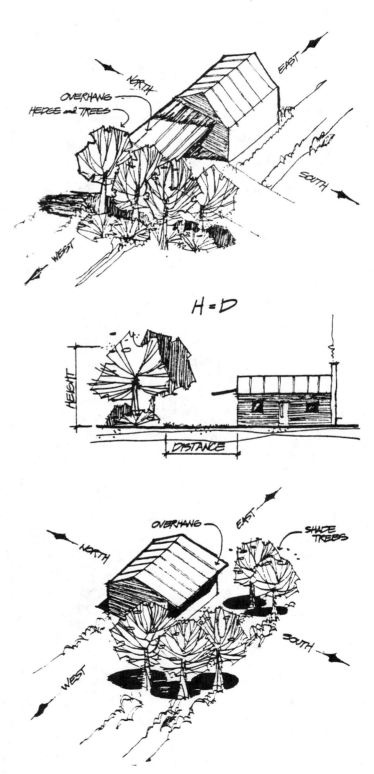

H = D

Western Sun

Western sun is the hottest summer sun with the most prolonged intensity during that part of the day when heat gain to a vertical surface is undesireable. The most effective shade for western summer sun is a vertical sun shade (Wright Oct. 1949 p. 158).

Overhangs are ineffective against western sun because the severe sun is afternoon sun, which comes in at a low angle. This sun requires a vertical wall to the west to screen it.

Sun Shield: Your best option is the use of a trellis sun shield with vine plantings. It is more effective than a sun shield wall of ordinary building materials since absorption and evaporation keep the vegetation and wall cool, as well as shading the structure (Wright Oct. 1949 p. 158).

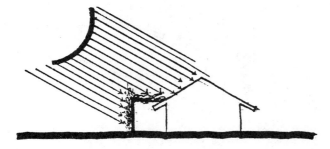

Vegetative Shading: Among the other options for vegetative shading for western exposures are tree plantings. The way in which west side tree plantings can provide shade for summer and permit sun in winter is illustrated in the following series (Langeweische p. 91).

■summer 3 p.m., shade wall of trees on western side tapering from tall to short shades west wall.

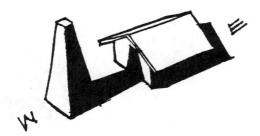

■summer 5 p.m. provided to front lawn, NW corner and most of the west side.

■winter 12 a.m., winter sun is not blocked, location of shade trees not selection of type (deciduous versus evergreen) makes the difference.

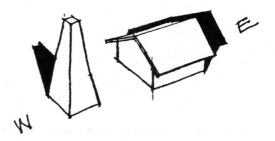

Location of structures: If you can't afford the time and money it takes for vegetative shading, consider the location of your garage. It can serve the same function as a buffer wall and shade summer western sun. Like the trees in the previous shading example, it won't block your winter afternoon sun either. Remember, a neighboring house can do the same (Langeweische July 1950 pp. 91-92).

Flexible Sun Shades: If these options aren't suitable, consider a flexible sun shade on your house. While they are an immediate investment in dollars, they are immediately completely effective. Among these options are:

■flexible shutters: provide view and block the sun

■awnings: awnings with long aprons block western sun
■sliding sun screens: provide shading to top and front of windows and wall as well

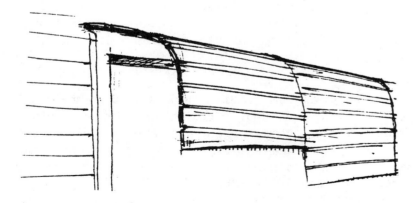

■ roll up porch shades on trellis frame blocks front of
  window but also permits ventilation
  (Langeweische March 1950 p. 131)

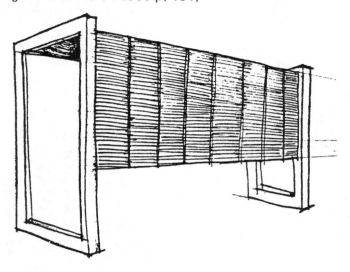

Eastern Sun

Eastern sun is as intense as western summer sun; however, its effects are less extreme because its incidence does not
coincide with the usual daily maximum air temperatures.
Shading response for eastern exposures utilizes the same techniques as those identified for western exposures.

Southern Sun

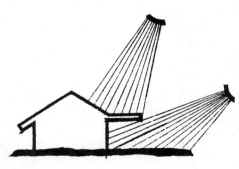

Southern summer sun is less intense on south-oriented
vertical surfaces than an eastern summer sun is on an east-
oriented vertical surface. During the winter, southern sun
penetrates structures easily due to its low angle. However,
during the summer there is a 45 degree increase in the southern
sun angle so that it can be stopped with an overhang. An
overhang is 100 percent effective in blocking southern summer
sun and will not interfere with winter sun penetration and
heating gain.

Your southern shading options might also include
vegetative shade plantings in conjunction with an overhang.
Some vegetation absorbs radiation and also cools due to evaporation. It is particularly suited for a south side where a paved
terrace or patio is planned. The following planting options
should be considered (Siple Nov. 1949 p. 159).
■ deciduous shade trees spaced at a distance equal to
  their height will give summer shade and let in winter
  sun.

■ trellis with deciduous vines gives shade in summer and lets in winter sun. It can also be extended to create a patio breezeway.

■ wood lath, nailed to quarter rounds of wood, makes a permanent eyebrow if overhang is not wide enough (Langeweische March 1950 p. 135)..

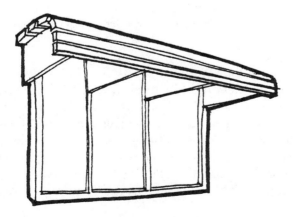

■ventilated metal awnings, that detach for winter, can shade windows and make a shower-proof terrace.

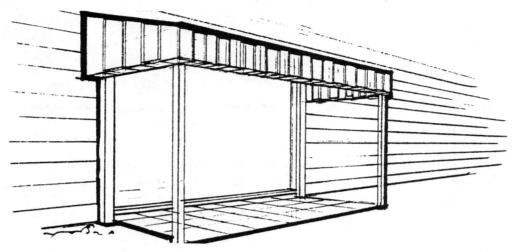

Measurement of Energy Savings

■shade planting on ES & W reduce solar heat gain (NAHB 1974 p. 54).

■shading on E & W can reduce sun loads by 1/5 (Olgyay p. 54) and reduce indoor temperature by 8 percent (AAN).

■exterior shading devices can reduce solar heat gain up to 80 percent, if they provide complete shading and permit air circulation (Hastings & Crenshaw pp. 2-16).

■exterior shading is 35 percent more effective than indoor shading because interior shading devices absorb, convert and re-radiate some of the radiation, striking them back into the interior of the structure (Olgyay pp. 69-71).

■shading glass windows can reduce heat impact by 1/3 (Olgyay p. 66).

■the shading on a roof or wall can reduce temperature of surface between 20 and 40 degrees.

Cooling

While we normally think of radiation effects in terms of daytime heating, the total cycle of radiation—incoming to the ground during the day, outgoing from the ground at night, is a consideration for overall cooling effects. The primary response to using radiation patterns for cooling is an architectural one using court-style housing as a 24 hour temperature control device. It was developed in hot arid climates and is primarily suited for hot arid regions of this country.

The climate principle on which this response is based is that the heat of the day and the coolness of the night both rise from the ground (Langeweische April 1950 pp. 205-207).

During the day, the sun heats the ground and the ground heats the air. Just before sunset, the earth gets cool and begins to cool the air (Langeweische April 1950 pp. 205-207).

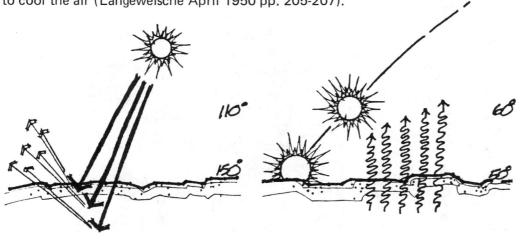

Use of court-style housing permits you to cut off radiation heat during the day and store cool nighttime air for coolness during the day.

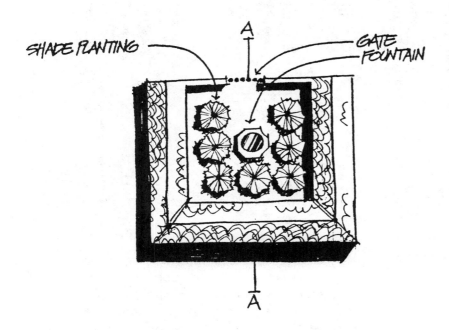

Basic components of the layout of court housing are court walls which enclose space, shade the dwelling unit and store evening coolness; open gate, which channels breezes; and water element, which provides evaporative cooling. Essentially, this is how the complex works during the day. For evening cooling, it is primarily the following principle of nighttime radiation which makes this layout desirable.

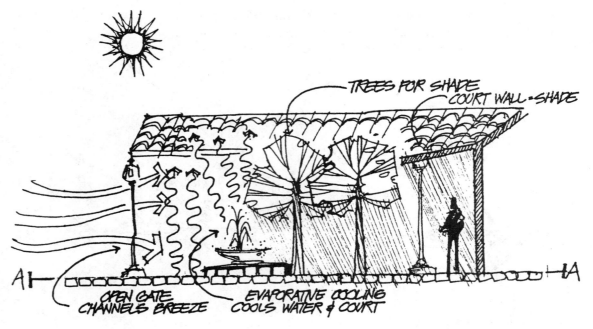

TREES FOR SHADE
COURT WALL-SHADE

A

OPEN GATE
CHANNELS BREEZE

EVAPORATIVE COOLING
COOLS WATER & COURT

A

Nighttime Heat Loss Varies With Direction: Heat loss from the ground at night occurs in three directions.

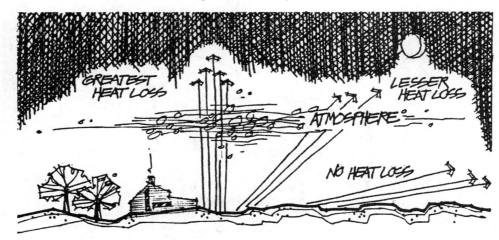

GREATEST HEAT LOSS

LESSER HEAT LOSS

ATMOSPHERE

NO HEAT LOSS

- perpendicular to the ground surface, which permits the greatest heat loss as radiation has to travel through the least amount of heavy atmosphere.
- oblique to the ground surface, which permits the least heat loss as radiation has to travel through the greatest amount of heavy atmosphere.
- almost level to the horizon, which traps hot air on the ground, permits no heat loss and often results in heat gain.

If we go back to your patio court layout, we can see that it permits maximum upward heat loss to the sky directly above—unlike an opened traditional structure which permits movement of evening heat in all directions (Langeweische April 1950 p. 205).

54

In considering this option, consider cost, time and effect. Also consider that on a larger scale the same principle holds true for planning and layout of structures. A multi-family cluster in the hot arid region could use the same principle.

Wind and Its Relationship to Heat Loss

The primary reason that wind control is important in reducing heating requirements is due to its effect on heat loss in a structure. The principle effects of heat loss can be summarized as follows.

Heat Transmission: is the flow of hot air from the interior to the exterior of a structure which results in the loss of 64 percent of interior heat.

Cold Air Infiltration: is the flow of cold exterior air to the inside of a structure which results in a loss of 15 percent of your interior heat (NSF p. 98).

**WIND CONTROL
TO PREVENT HEAT LOSS**

HEAT TRANSMISSION
OUTWARD

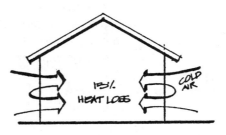

COLD AIR INFILTRATION
INWARD

This significance of wind in terms of heat loss is due to its velocity. The following effects of wind exposure on heat loss and energy consumption have been identified.

Measurement of Heat Loading Effects
The following effects of wind on heating load of structures have been identified.
- a 20 mph wind can double the heating load of a building (Fitch p. 104).
- heating of an unprotected house is 4 times heating load of a protected house (Olgyay pp. 98-99).

Measurement of Energy Consumption Effects
The following effects of wind exposure on energy consumption for heating have been identified.

Princeton: Townhouse Experiment
- townhouses exposed on the windward side use 5 percent more gas than houses on the leeward side sheltered by other buildings.
- townhouses at the end of the row with an exposed wall use 10 percent more gas than other units (Harwood pp. 21-22).

South Dakota: House
- an electrically heated house, fully exposed to the wind, uses more energy for the entire winter to maintain an interior temperature of 70 degrees F. than an identical house sheltered by a windbreak (Hastings and Crenshaw pp. 1-5).

Measurement of Energy Conservation Effects
The following effects of wind protection on energy conservation have been identified.

Nebraska Test House: Experiment conducted on two identical test houses with one exposed to the wind and one protected. A 70 degrees F. constant interior temperature was maintained. The following fuel savings were identified:
- the protected house had a 22.9 percent fuel savings.
- with good protection on 3 sides, 30 percent fuel savings were estimated (Olgyay p. 99).

Principles of Wind Control

The primary characteristic of wind is velocity and flow pattern. In responding to wind by either attempting to buffer wind or channel breezes, the following principles must be considered.

Velocity: Wind cannot be stopped, it can only be slowed down and averted. An attempt to stop wind with a solid barrier does not work, instead it causes suction pressure and swirling on the windward side of the wall forcing the wind to break over the top and whirlpool. This effect is magnified with the height of the barrier, the most commonly seen example being the wind pattern around tall buildings in cities (Langeweische Oct. 1949 p. 164).

Flow Pattern: Wind cannot be stopped, it must have a place to go. To deflect wind, but still permit a flow pattern to be maintained a surface slant-wise to the wind is better than one set squarely against it. A good example of this is the slant of a pitched roof set against the wind which provides a better deflection and flow director than a wall (Langeweische June 1950 p. 91).

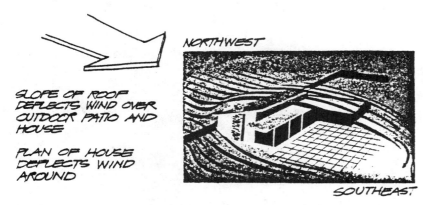

SLOPE OF ROOF DEFLECTS WIND OVER OUTDOOR PATIO AND HOUSE

PLAN OF HOUSE DEFLECTS WIND AROUND

NORTHWEST

SOUTHEAST

Options For Wind Control

Because the velocity of wind is correlated with winter heating requirements, wind barriers are often required. Even if you have selected the optimum site or orientation, or if your present one is less than optimum, you might consider one of several wind barrier techniques. Keep in mind that different techniques are suited to different levels of planning and design.

Large Scale Control: To direct and deflect wind involves siting, orientation and design of your house, trees and garage.

UUUUUUUU → HEAT LOSS

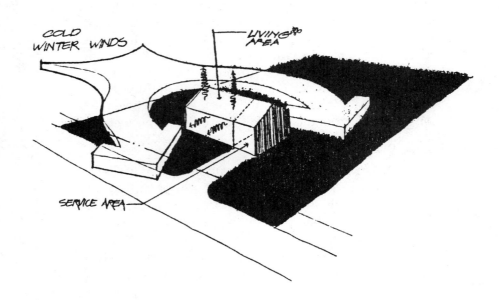

COLD WINTER WINDS

LIVING AREA

SERVICE AREA

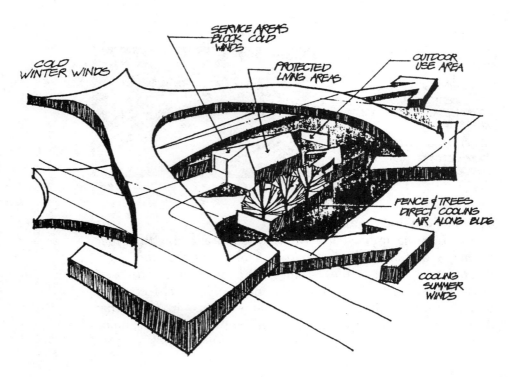

SERVICE AREAS BLOCK COLD WINDS

PROTECTED LIVING AREAS

OUTDOOR USE AREA

COLD WINTER WINDS

FENCE & TREES DIRECT COOLING AIR ALONG BLDG

COOLING SUMMER WINDS

Measurement of Orientations Effects on Air Infiltration

According to simulation studies of comparative effects of orientation and air infiltration conducted at Twin Rivers, New Jersey, the following correlation between building orientation and air infiltration for exposed town house units has been identified.

45° Angle Orientation to the wind is the worst possible orientation as it causes wind to flow towards the corner of structures, striking the front and side walls. Air infiltration measured for this unit orientation was 59% greater than an orientation with the front of units perpendicular to the wind. The 59 percent increase was measured for the end unit of the townhouse row (Mattingly and Peters 1975 p. 33).

Parallel Orientation of the sidewall to the wind produced the minimum amount of infiltration when the measured unit was either on the interior or lee side of a townhouse row. However, when measurements were taken at the end unit exposed directly to the wind, air infiltration was recorded as 52 percent greater than when the front wall of unit was hit directly (Mattingly and Peters 1975 p. 37).

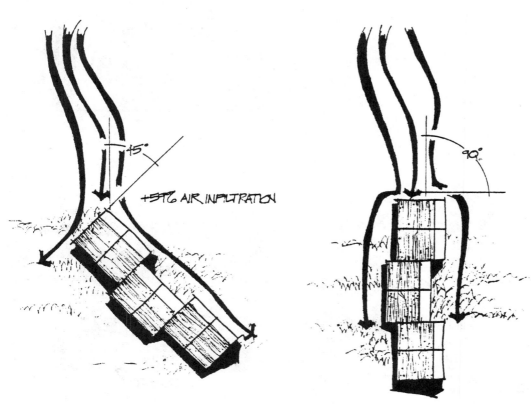

Measurement of Buffering Effects on Air Infiltration

According to the Twin Rivers, New Jersey, study already cited (Mattingly and Peters), use of solid fencing and evergreen vegetation on the windward side of town houses resulted in the following reductions in air infiltration.

Solid Fence: 5'6'' at 6'' off the ground on the windward side reduced air infiltration by 26% for end wall units and 31% for interior units (Mattingly and Peters 1975 p. 37).

Single Row of Evergreens: At a height equal to the height of the unit located at a distance equal to 1½-2 times the height of a unit spaced to permit boughs at the base to touch resulted in a 40% air infiltration (Mattingly and Peters 1975 pp. 37-38).

Combined Fence and Evergreens: On windward side of town houses resulted in a 60% reduction in air infiltration (Mattingly and Peters 1975 p. 38).

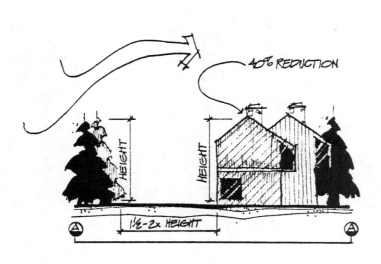

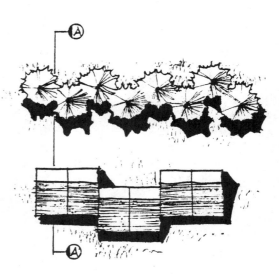

Small Scale Control: To stop spillage, cross flow and whirlpool effects involves siting and design of walls, fences, hedge.

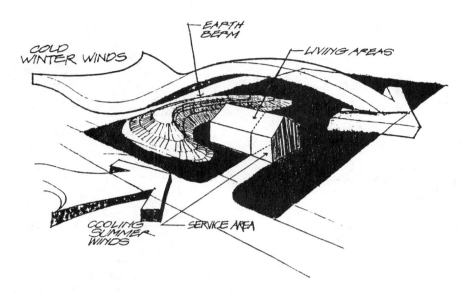

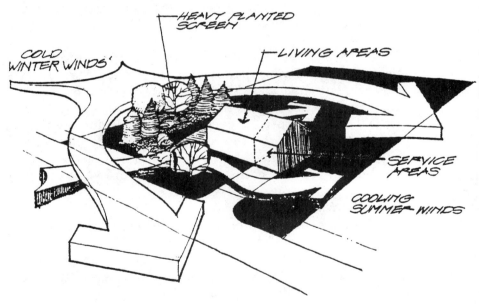

In buffering wind, remember that it should be *deflected* not stopped. Also, it should be deflected so that it has a place to go. Through proper design of wind barriers you can determine where the wind will go (Langeweische June 1950 p. 91).

Essentially, there are two types of control—deflectors and barriers. Here is how they work.

Wind Deflectors: They divert air flow, but do not stop wind. They scoop up oncoming air, bend it and make it flow higher.

Wind Barriers: Effective wind barriers are wind penetrable barriers, they do not stop wind but slow it down due to frictional drag. The most effective wind barriers of this type are man-made barriers based on a half-slot snow fence design or natural tree barriers whose branching and twigs permit the same type of through flow (Langeweische June 1950 p. 97).

Though there are regional and site specific variations, the general cold wind pattern for the United States is NW wind. In buffering and directing wind with any device, first determine the site specific pattern of prevailing cold wind and respond to that specific condition. Options for controlling cold winter wind are shown in the following illustrations. In selecting any option, consider the needs versus cost, time and degree of effectiveness.

Vegetative Windbreaks

In using vegetation as a windbreak, the following design criteria should be considered.

Placement: Shelterbelts and windbreaks are most effective when they are placed perpendicular to the prevailing wind.

PREVAILING WIND

Height: Height of trees is probably the most important characteristic, because the distance that protection extends to leeward is proportional to the height of the windbreak. The distance that protection extends is therefore commonly expressed in windbreak heights.

For example, the amount that wind is reduced will differ at leeward distances of 4, 10 or 20 times the average height of barrier, but the percent reduction at 4 H, for example, is the same regardless of barrier height.

Density: Density of different height levels in windbreaks is important for wind reduction. It is also a consideration in terms of the amount of time required for effective wind protection. The following series illustrates density levels and comparative effectiveness over time (Read pp. 11, 12).

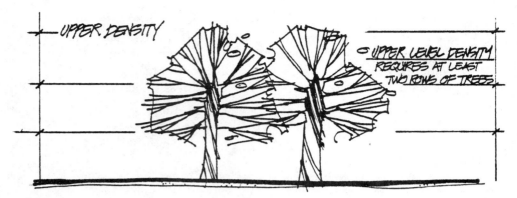

Upper level density requires at least two rows of tall trees which are effective for the first 15 years of growth.

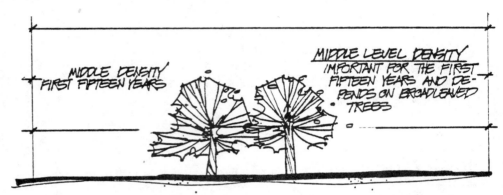

Middle level density is also important for the first 15 years of growth and requires low, broad-leaved trees.

Permanent middle level density depends on pines and red cedars for lifetime effectiveness.

Lower level density for the first 15 years is provided by all three forms of vegetation as they grow.

After 20 or 30 years, most conifers and broad-leaf trees lose their leaves, and permanent density must be provided by low thick shrubs.

Length and Width: Length and width of windbreaks is only important in so far as it effects density. Making windbreaks wider than necessary to achieve moderate density has no added effect on wind reduction. In general windbreaks patterned on multiple rows produce maximum desired foliage density at all levels (Read p. 15).

Configuration: The configuration of windbreak affects the degree of wind reduction and penetration. Shelter-belts with a pitched roof cross-section are the least effective at stopping wind and should be avoided.

Preferred configuration is one with a strong vertical outline.

The most effective windbreak configuration is an irregular one which is more effective in reducing eddying than a uniform one.

Also, a mixture of species and sizes producing a rough upper surface is more effective in controlling wind (ASLA p. 35).

Measurement of Climate Modification

Windbreaks can produce velocity by 40% over horizontal distances up to 4 times their height. Wind at 20 mph can be slowed to 10 mph, and a 40 mph wind can be slowed to 20 mph. General effect is in proportion to height and distance as illustrated (Read p. 3).

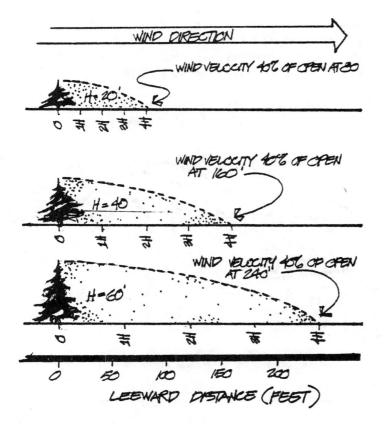

WIND DIRECTION

WIND VELOCITY 40% OF OPEN AT 80'

H=20'

0 ½H 2H 3H 4H

WIND VELOCITY 40% OF OPEN AT 160'

H=40'

0 1H 2H 3H 4H

WIND VELOCITY 40% OF OPEN AT 240'

H=60'

0 1H 2H 3H 4H

0   50   100   150   200

LEEWARD DISTANCE (FEET)

*Measurement of Energy Savings

1. Wind barriers of trees, vines and plantings on the south and west side can reduce winter fuel consumption in range from 10 to 40 percent. (AAN)

2. Formula for Quantification (Olgyay p. 98.)

    heating load/wind velocity/temperature difference/ shelter belt location

    $$\frac{Q}{T} = 1.3 \ (10^{0.018L0.7u})$$

    Q = heating load in BTU per hour

    T = difference between inside and outside temperature in degrees Farenheit

    L = distance from belt to house in barrier heights

    u = wind velocity in mph

Wind Fences

Since the chilling effect of wind is proportionate to its speed, wind fences which decrease velocity can be effective in creating warm zones on the lee side (Howland June 1950 pp. 101-102).

Solid barriers are ineffective windbreaks as they try to stop wind and create problems of eddying and swirling. Don't use them.

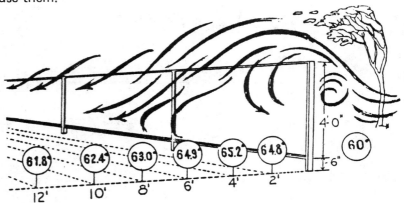

Solid barriers with sloping baffles are effective in creating a warm zone and decreasing velocity because the baffle directs wind up in a gentle arc and avoids downward, wave-like whirlpooling.

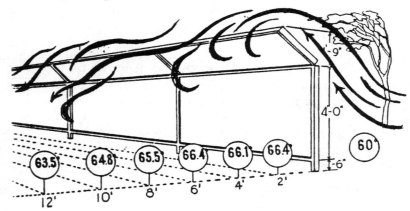

Solid barriers with slanted baffle toward the wind get a somewhat smaller protected area, but note that close in lee of fence, you get highest temperature readings in any fence tested: 67.5 degrees F. Even at three times its height behind fence, protection was noticeable.

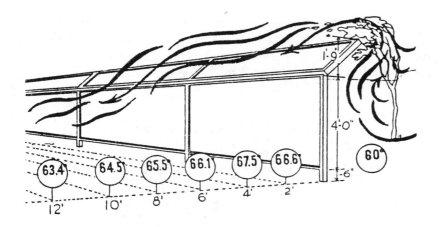

Horizontal louvers tilted up away from wind, proved better, all-around windbreak than solid fence. Temperatures were not quite as high but protected area was wider. This fence would be good where wind velocities had to be reduced but ventilation kept.

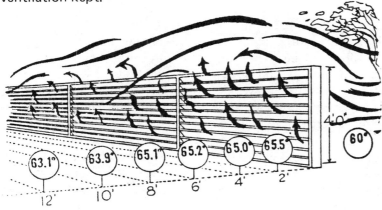

Tilting the louvers down makes the least effective windbreak of all fences tested because the louvers bounce the wind right down into the area to be protected. This is not a warm fence.

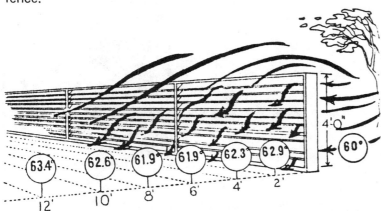

Vertical laths, spaced a half inch apart, made a good windbreak. Although temperatures were fairly low in immediate lee, they rose steadily as you moved away. *At a distance of three times its height, this fence gave best protection of all.*

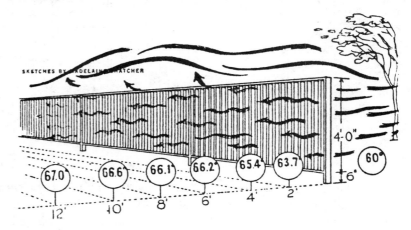

Walls

Walls are another device which can be used to modify cold wind and air. They are most suited to climate regions or sites where frost and severe cold night air is a problem. As with any other of the wind devices, they work on the principle of channelling cold air flow and draining frost—they do not attempt to stop it, merely to direct it.

At night, cold air creeps along the ground and brings frost to low places where it collects. If you have a site with cold air pockets, you might consider a device such as the wall illustrated below. It works by daming, draining and collecting cold air.

The wall is designed to calm cold air on the uphill side with a wall opening on the downhill side to act as an outlet. The house plan and evergreen divert air from the center of the site and the house. The design is based on the pattern of cold air flows down hill at ground level and encourages the natural pattern to eliminate frost pockets (Langeweische June 1950 p. 204).

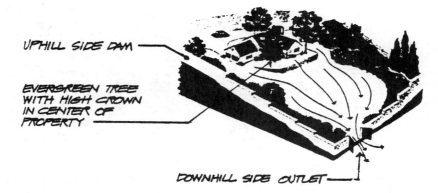

UPHILL SIDE DAM

EVERGREEN TREE WITH HIGH CROWN IN CENTER OF PROPERTY

DOWNHILL SIDE OUTLET

Subterranean Housing

While this technique is primarily suited for areas that are not humid, it is an immediate and effective way to insure minimum heat loss in a structure. Subterranean housing is based on the following principles (NSF July 1975 pp. 89-92).

Elimination of Heat Loss: Conventional structures lose heat in two ways. One is by heat transmission through walls, the other is air infiltration into the building. Since underground houses have a minimum surface exposure to the cooling effect of air and wind, infiltration losses are greatly reduced.

Soil Temperature: Soil temperatures are very slow in change. About the time that you start to need to heat your house the soil is reaching its maximum temperature from the summer heat. This thermal lag reduces heating and cooling needs.

Basic requirements for subterranean housing are as follows:
- insolation, waterproofing and dehumidification
- low water table
- subsurface drainage
- detailed soils analysis of your site

3' of earth and trees on concrete structure, waterproofed air space all around insulated living envelope.

Measurement of Energy Savings

In studies done at 25 degrees F. it was determined that heat loss through a wall with eight inches of insulation is 6.5 times greater than that through an uninsulated underground concrete wall.
- heating needs can be reduced by up to 75 percent
- even a thin layer of soil on the roof of a house and a berm around it can reduce energy usage by 30 percent (Smay April 1977 pp. 85-91).

Earth Berming

If site conditions are unsuitable, or the concept seems too radical, earthberming might be considered as another option for reducing heat loss. By using the earth berming on the windward side and orienting major windows for optimum sun, measurable heating savings can be achieved.

Three feet earth berming to window sills reduced heat loss 6.6 percent for a rectangular house and 13.1 percent for a square house (Kroner and Haviland pp. 30-31).

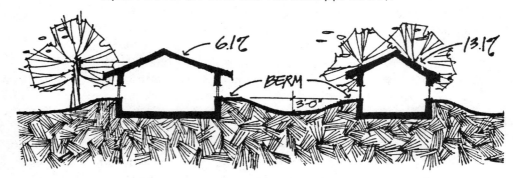

Earth berming two feet below roof eave, leaving full length door and window areas, provided a 31.7 percent reduction in heat loss.

WIND CONTROL FOR COOLING

We've looked at techniques to modify winter wind, now let's look at techniques to use summer breezes to your advantage.

The positive role of wind which you want to maximize is the cooling effect of summer breezes. Basically, the hot weather wind requirement is to pick up enough wind to dry perspiration. Too little wind makes you feel damp and hot. Too much wind makes you feel dry and hot.

The big factor affecting response to summer wind is the degree of humidity in a climate region. Humidity or mugginess is due to vapor pressure which prevails during July and August, and which keeps perspiration from evaporating.

Regional requirements for summer wind correlated with humidity are illustrated on the map below:

WIND
REQUIRED
FOR
COMFORT

— less than 1 m.p.h.

— 1 m.p.h.

— 3 m.p.h.

— 5 m.p.h.

— 7 m.p.h.

— over 7 m.p.h.

Variable wind requirements are as high as 7 mph in the hot humid zone (Florida, Georgia coastal regions) and as low as 1 mph in the temperate zone (Kansas City, Chicago and New York City) (Langeweische June 1950 p. 125).

Basic principles of climate which relate to summer cooling are as follows:

Day and Night Temperature Distribution: As the discussion of radiation pointed out, daytime radiation is incoming radiation to the ground. Temperature distribution measured along the vertical height of a two-story house illustrates how significant the differences are: during the day, temperature is hottest at the ground level and 9 degrees cooler (almost 10 percent) at the roof line (Langeweische June 1950 p. 165).

The air you want to capture is the coolest available—lower ground air at night, upper level air during the day. Remember, however, that air above 85 degrees cannot cool and if carried by a breeze it can even make you hotter, so only encourage summer air flow of cooler air. Your main response will be to capture evening breezes as close to the ground as possible. Most of these techniques relate to the layout and design of your structure.

Air Intakes: These can be 100 percent effective at capturing cool air and circulating it if they are placed at the right level. Remember the comparative day and night temperatures near the ground for air intakes should be taken at ground level. Locate them there and don't block them with vegetation or earth (Langeweische June 1950 p. 124).

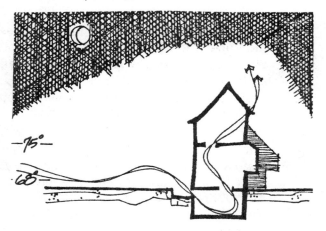

Shade Fences: If your daytime temperature is cool enough and winds aren't hot, remember your windbreak or sun shade can also be used to channel breezes and ventilate. The proper combination of an open slat system with vegetation to cause evaporative cooling can make your sun shade a breezeway as well (Langeweische June 1850 p. 124).

2

regional
site
design
guidelines

cool region

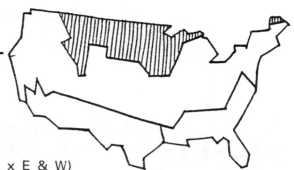

CLIMATE CHARACTERISTICS

Winter
- Sunlight probability 0.35 (BSIC p.20)
- Maximum radiation on S vertical walls (2 x E & W)
- Prevailing wind NW & SE

Summer
- Sunlight probability 0.60
- Maximum radiation on E & W vertical walls (2½ x S)
- Prevailing wind NW & SE

DESIGN CRITERIA

Winter
- Heating requirements: 7,000 degree days (BSIC p.23)
- Increase heat production, radiation absorption
- Decrease radiation loss
- Winter response is 3 times as important as summer response (Olgyay p. 155)

Summer
- Reduce conduction and evaporation loss (Olgyay p. 155)

GROSS SITE SELECTION GUIDELINES

Look for sites with slope and orientation to increase day time winter radiation and decrease winter winds. Look for:

Landform
- Lee side of hill
- Any location out of prevailing NW winter wind
- Avoid valley bottoms

Water
- In Great Lakes region, shoreline locations have the advantage of:
  - heavier cloud cover
  - extended autumn and spring season
  - shoreline breezes
- In Great Lakes region, inland locations have:
  - greater temperature extremes
  - lighter snow blanket

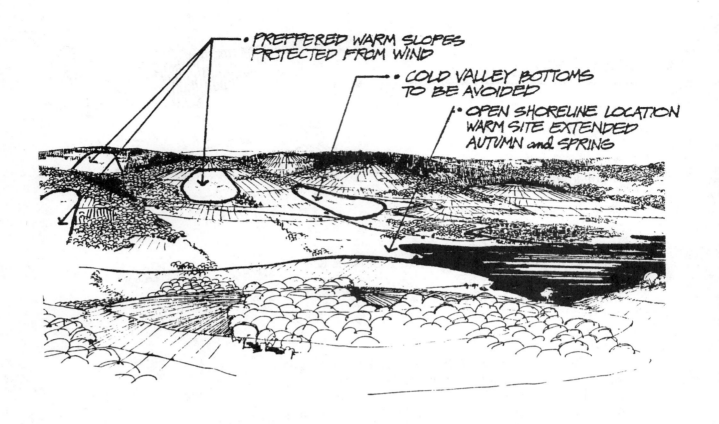

- PREFFERED WARM SLOPES PROTECTED FROM WIND
- COLD VALLEY BOTTOMS TO BE AVOIDED
- OPEN SHORELINE LOCATION WARM SITE EXTENDED AUTUMN and SPRING

DISCRETE SITE SELECTION GUIDELINES

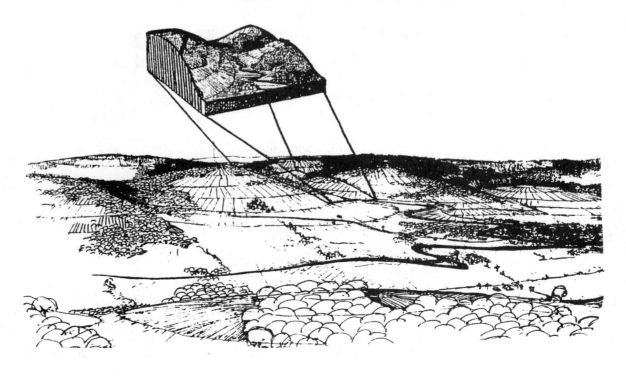

Site Analysis

Analysis and selection of optimum residential sites would be based on the following general criteria:

- Identify and select sites with the greatest protection from prevailing winter winds as cool region has the highest winter heating requirement (7,000 degree days) of any part of the country.
- Identify and select sites with warm slope orientations South or Southeast preferred.
- The relationship of topography and vegetation to winter wind and sun are the primary determinants for site selection.

Slope Direction and Gradient

- Select South-Southeast slopes for residential development
- Avoid high water table due to humidity and frost
- Avoid gradients in 10%-20% due to susceptibility to erosion, cost of development and potential of disturbing surface air flow and snow drifting patterns

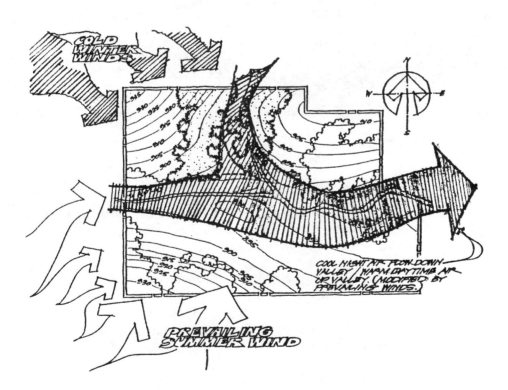

Seasonal Wind and Cold Air Flow
- Avoid ridge locations exposed to winter wind
- Avoid valley locations which are cold, damp potential frost pockets
- Select SE slopes with exposure to summer breezes

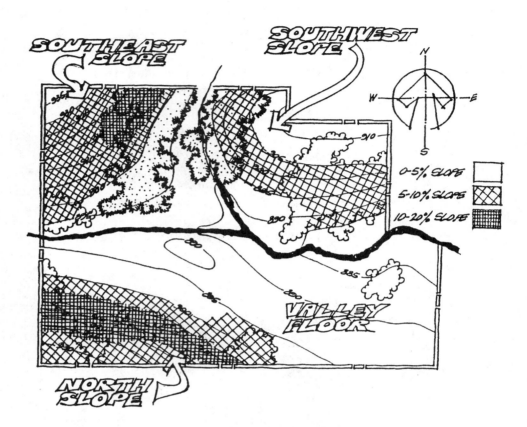

Site Selection

Site selection should be based on combined consideration of winter and summer climate influences related to existing landform, vegetation and natural features.

Development Suitability: for various uses, structural types and use intensities.

Recommended guidelines for various uses and structural types are as follows:

Residential Structures: Select S-SE slopes for winter sun gain

Tall Structures: Select sites out of prevailing wind and downward from lowest elements (ASLA p. 183)

Exterior Open Space: Select open sites with periodic shading sheltered from NW wind

Exterior Living Space: Select sites on S side of dwelling for protection from NW and N wind and exposure to summer and winter sun

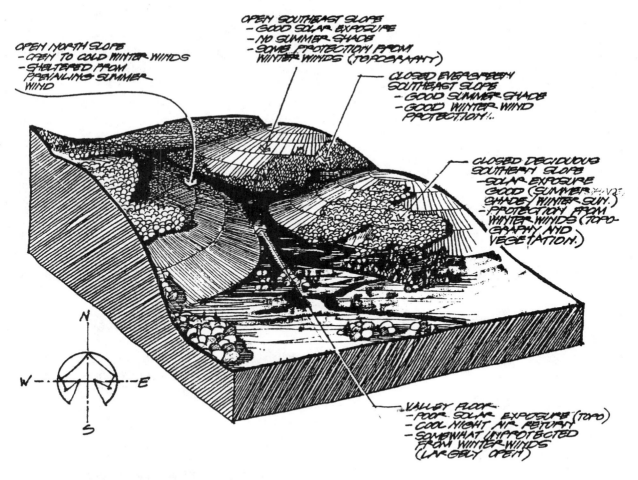

OPEN NORTH SLOPE
- OPEN TO COLD WINTER WINDS
- SHELTERED FROM PREVAILING SUMMER WIND

OPEN SOUTHEAST SLOPE
- GOOD SOLAR EXPOSURE
- NO SUMMER SHADE
- SOME PROTECTION FROM WINTER WINDS (TOPOGRAPHY)

CLOSED EVERGREEN SOUTHEAST SLOPE
- GOOD SUMMER SHADE
- GOOD WINTER WIND PROTECTION

CLOSED DECIDUOUS SOUTHERN SLOPES
- SOLAR EXPOSURE GOOD (SUMMER SHADE / WINTER SUN.)
- PROTECTION FROM WINTER WINDS (TOPO-GRAPHY AND VEGETATION.)

VALLEY FLOOR
- POOR SOLAR EXPOSURE (TOPO)
- COOL NIGHT AIR RETURN
- SOMEWHAT UNPROTECTED FROM WINTER WINDS (LARGELY OPEN)

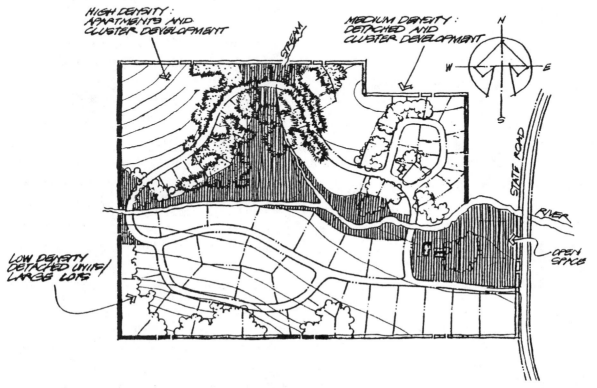

HIGH DENSITY: APARTMENTS AND CLUSTER DEVELOPMENT

STREAM

MEDIUM DENSITY: DETACHED AND CLUSTER DEVELOPMENT

STATE ROAD

RIVER

OPEN SPACE

LOW DENSITY DETACHED UNITS/ LARGE LOTS

Development Suitability: for various use intensity should be based on combined consideration of landscape type, slope orientation and gradient, presence of dense mature vegetation and site accessibility.

Guidelines for various use intensities correlated with landscape types have been identified as follows (Kaminsky p. 35*):

High Intensity Use: Warm slopes (S, SE, SW, E) and flat or rolling land (0%-8%) with high accessibility
Moderate Intensity Use: Warm slopes over 8% with high accessibility
Low Intensity Use (Low Building Coverage): Cold slopes (N, NE, NW) with extensive tree cover
Semi Rural Intensity Use (Low Intensity and Low Building Coverage): Slopes in excess of 15%
Existing Regional and Site Specific Energy Consumption Patterns as indicated by existing developments.

An independent study conducted jointly by the Reimann-Buechner-Crandall Partnership, Landscape Architects, and the Niagara Mohawk Power Corporation of Central New York was undertaken during July of 1977. The study attempted to correlate energy consumption for heating and the following site characteristics:

Topography: Slope gradient and orientation
Vegetation: Type and degree of coverage

The site factors identified above were used as comparative variable factors. The following were held constant:

Housing Characteristics: Construction, age, size
Development characteristics: Denisty, housing type

The study methodology was to select specific residential development sites in Syracuse, New York, with variation in either topography or vegetation characteristics. Energy consumption data for space heating of houses on these sites has been provided by the Niagara Mohawk Corporation and used as the basis for comparing energy efficiency.
Study conclusions and sites illustrated on the following pages indicated the following:

Highest Energy Efficiency Related to Slope Orientation: is found on south or southwest slope developments.

NOTE: Classification system and landscape types developed for a specific study site in Baltimore County but are suited for requirements in this region. Conceptual framework of the system which incorporates Floor Area Ratio (FAR) and trip distance, suggests adaptability to various regions of the country.

Further findings of the Mattingly and Peters study indicate that front wall exposures perpendicular to the wind increase wall exposure and wind infiltration.

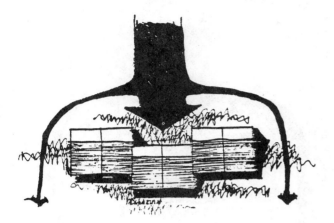

Worse still is a 45° angle tilt to the wind which will increase air infiltration by 59% over the perpendicular orientation due to exposure of both side wall and front wall.

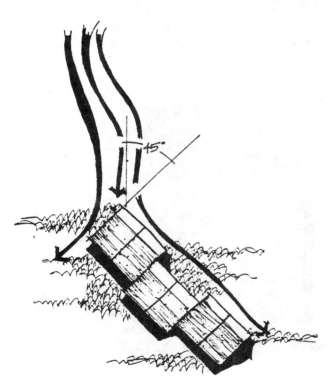

If forced to use a front wall exposure perpendicular to the wind consider use of fencing on the windward side for a 26% reduction in air infiltration (Mattingly & Peters p. 37).

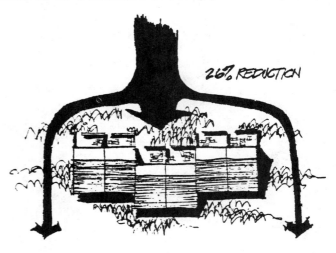

Consider a straight line of evergreen trees on the windward side located at a distance 1½–2x the height of the unit and spaced so that bottom boughs touch. This has been shown to result in a 40% reduction in air infiltration versus an unsheltered situation (Mattingly & Peters pp. 34, 56).

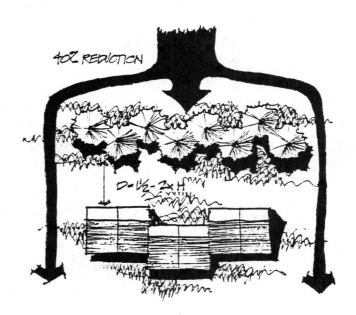

Consider combining hedge wood fence combinations for a 42% reduction in air infiltration versus an unsheltered situation (Mattingly & Peters pp. 35,36).

Essentially, all of these responses are patterned on parallel air flow orientation which is the recommended optimum.

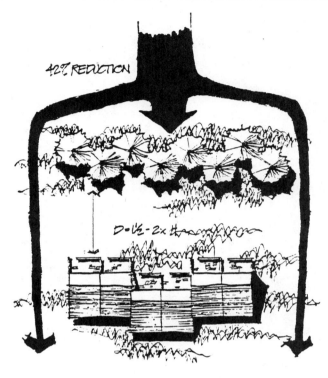

Sun Response: orientation and exposure

Sunlight probability for the cool region of the country is the lowest of all regions. While utilization of solar heating devices does not seem likely, building relationships should still be considered in terms of maximizing winter sun.

Orientation 12° south-southeast on slopes of 0-20% is the optimum. Even though winter sun is not predominant in this region, preliminary indications are that south-southwest orientations can be up to 10% more energy efficient for winter heating (Reimann-Buechner-Crandall/Niagara Mohawk).

Also in winter, south facing vertical walls receive twice as much solar radiation as east or west.

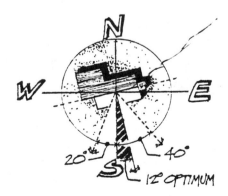

The following guidelines are recommended for optimizing winter sun.

Avoid placing taller building between sun and lower structures. Place highest elements down sun from lowest elements to minimize winter shading (ASLA p. 183).

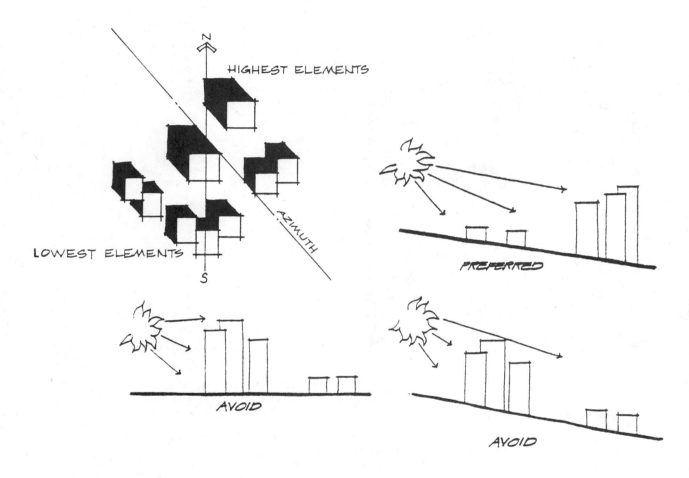

Place structures halfway up S SE slopes using middle to lower middle of slope to prevent excess winds effects and cool air pockets (Olgyay p. 155).

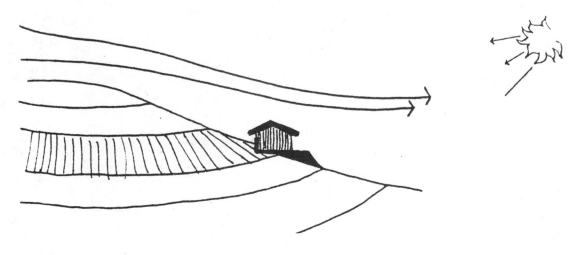

Cluster multiple dwellings around sunny courts to create sun pockets.

On detached dwelling units, utilize exterior walls and fences to capture the winter sun and reflect warmth into activity areas.

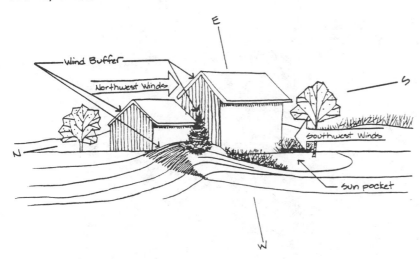

Structures can be built into hillside or partially covered with earth and planting for natural insulation.

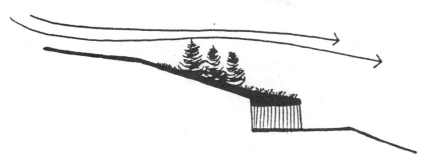

Place blank walls, garages and storage areas on the north or northwest side of dwellings to buffer winter winds.

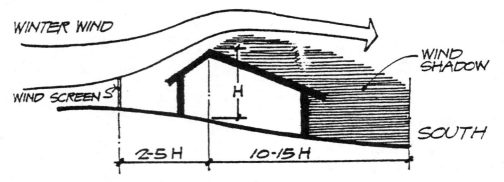

Clustered Attached Dwellings

For clustered multi-family dwellings, terraces and outdoor living areas should be integrated within the building clusters. This will reduce cold air movement in winter and will channel and direct breezes in summer.

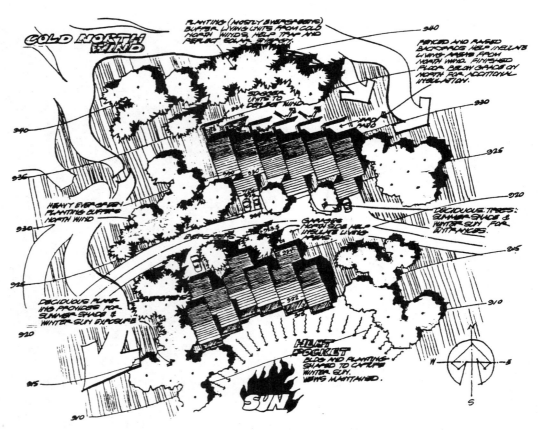

Streets and parking areas shaded with deciduous vegetation will also channel summer breezes and reduce radiation reflection while allowing the sun to penetrate during the winter.

Cluster buildings to create sun pockets and take full advantage of winter sun.

Use solid walls and single row hedge planting on end units facing north westerly wind pattern.

Locate fences on northside and evergreen wind break at 1½—2x the distance of the height of the unit. Use vegetation equal to the height of the unit and spaced so that bottom boughs touch.

Wind barrier and blank westerly wall will also isolate undesired summer westerly sun.

Also consider compacted layouts around interior court combining end walls, fences and vegetation for wind protection.

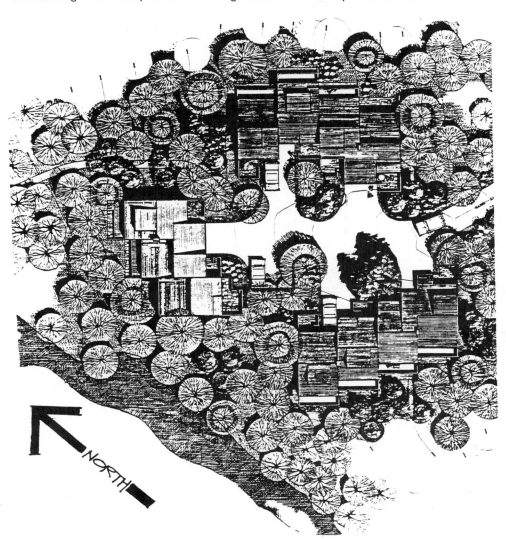

Single Family Detached Dwellings

Space structures and circulation systems to permit maximum radiation, buffer winds and channel breezes.

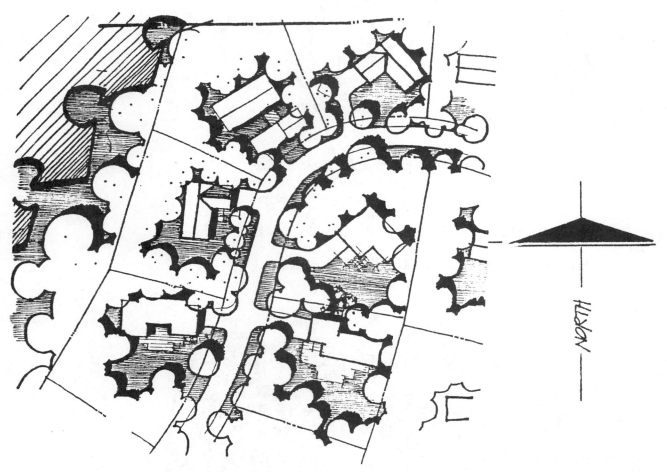

Orient buildings on EW axis.

Minimize and insolate wall surfaces exposed to wind.

Locate garage and service structures as wind buffers to channel wind flow and snow drifting.

Underground or semi-subterranean site integration is suitable if soil is suitable and region is not humid (see Wind Control, Section I, Part II).

temperate region

CLIMATE CHARACTERISTICS

Winter
- Sunlight probability 0.45 (BSIC p.21)
- Maximum solar radiation on S vertical walls (2 x E&W)
- Major wind pattern northwest

Summer
- Maximum solar radiation on E & W vertical walls (2½ x S)
- Major wind pattern SSW

DESIGN CRITERIA

Winter
- Heating season 5,000 degree days (BSIC p. 21)
- Maximize solar radiation and heat gain
- Minimize northwest exposure and heat loss 60% of the year (Siple Nov. 1949 p. 203)

Summer
- Cooling requirements 479 hours (BSIC p. 24)
- Maximize breezes and exterior living area, 31% of the year (Siple Nov. 1949 p. 203)

Year Round
- Maintain seasonal balance of heat production, radiation and convection (Olgyay p. 161).

GROSS SITE SELECTION GUIDELINES

Look for sites with optimum combination of winter radiation and summer breezes. Look for:

Landform
- Sloping sites with south southeast direction preferred, will be warmest.
- Avoid valleys which are cold, damp and potential frost pockets.

Water
- Proximity to large water bodies is preferred for summer cooling effect.
- Avoid areas of coastal fog.
- In areas of fog, look for higher elevations where fog retention is less.

■Avoid sites with N NE wind exposure especially where topography is steep and open; they will be extremely cold in winter.

■Look for sites with S SE winds, they are warmer in winter and cooler in summer.

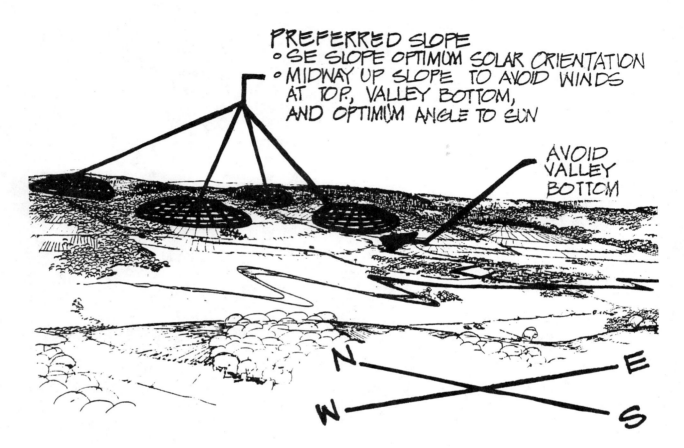

PREFERRED SLOPE
○ SE SLOPE OPTIMUM SOLAR ORIENTATION
○ MIDWAY UP SLOPE TO AVOID WINDS AT TOP, VALLEY BOTTOM, AND OPTIMUM ANGLE TO SUN

AVOID VALLEY BOTTOM

DISCRETE SITE SELECTION GUIDELINES

Site Analysis

Analysis and selection of optimum residential sites should be based on the following general criteria.

Identify and select sites with the greatest balance between winter heating requirements due to wind exposure and summer cooling requirements due to sun exposure. This region falls in the middle in terms of heating and cooling requirements.

Identify and select slopes with warm orientations (S-SE). Avoid western slopes due to summer cooling requirements.

The relationship of topography, vegetation and proximity to water to sun and wind in both winter and summer are the primary determinants for site selection.

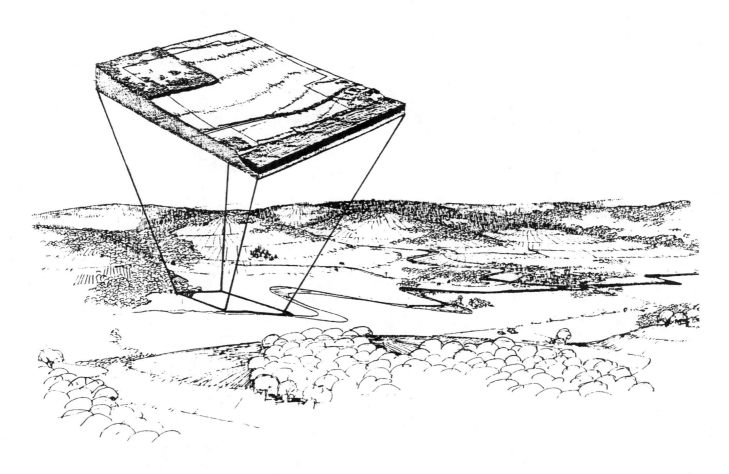

Summer Breezes and Winter Wind Patterns
- Avoid sites which are opened to N NW winter winds.
- Select sites which are opened to S SE summer breezes.
- Avoid valley bottoms and low spots which are potential cold air dams and frost pockets.

Slope Direction and Gradient
- Select S SE slopes for residential development for greatest balance of summer and winter sun and wind.
- Select sites on middle slope, within thermal belt, rather than at the foot or crest of a slope, for maximum radiation. In general, sites on upper or middle slope receive maximum winter radiation and less summer radiation than horizontal sites at the foot or crest of a slope.

NORTHWEST WIND

LEESIDE THERMAL BELT BELOW CREST

■In valley situations the thermal belt is preferred unless the slope is exposed to winds, then avoid crests and locate halfway up the slope (ASLA p. 172).

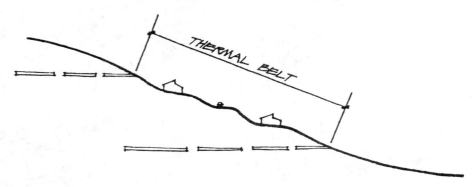

Site Selection

Site selection should be based on combined consideration of:

Winter and Summer Climate Influences: related to existing landform vegetation and natural features

Planning and design studies conducted by the firm of Rahenkamp, Sacks, Wells and Associates in metropolitan New York and New Jersey indicate the following guidelines for site selection and use.

In a valley situation (running Northeast to Southwest) with high water retention soils, the valley was eliminated for residential development and maintained as a natural drainage system.

Vehicular circulation systems were sited diagonally across the valley and used as dams for natural holding ponds.

Valley sites were permitted to maintain natural vegetated character and serve as shelter belt in the winter and cross ventilating natural air conditioner parallel to summer breezes for cooling (Goldberg, unpublished).

Development Suitability: Climate and design studies by Olgyay recommend the following guidelines for various uses and structural types.

Residential Structures: Select S-SE warm slopes.
Large Structures: Locate high buildings out of prevailing winds and venture effects.
Exterior Open Space: Locate in open shaded areas.
Exterior Living Space: Utilize S-SE orientation for patios and sun pockets.

DESIGN GUIDELINES
Plan Layout

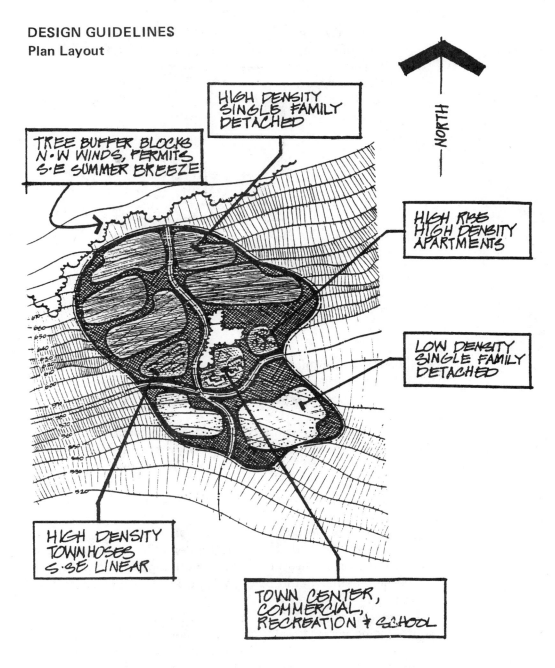

HIGH DENSITY
SINGLE FAMILY
DETACHED

TREE BUFFER BLOCKS
N·W WINDS, PERMITS
S·E SUMMER BREEZE

NORTH

HIGH RISE
HIGH DENSITY
APARTMENTS

LOW DENSITY
SINGLE FAMILY
DETACHED

HIGH DENSITY
TOWNHOSES
S·SE LINEAR

TOWN CENTER,
COMMERCIAL,
RECREATION & SCHOOL

Guidelines for plan layout are as follows (Olgyay p. 160-165).

High Density Single Family Clusters: are sited on preferred south east slope in loose arrangements to permit free air movement and solar penetration.

Low Density Single Family Detached: are sited on northeast slopes with short axis oriented toward wind.

High Density Apartments: are sited on flat area near water for ventilation with vegetative wind break.

Townhouse Units: are sited on flat area with blank short end oriented to wind and balcony arrangements for exterior southern clearing space.

Town Center and Commercial Facilities: are sited in flat area central to all development and connected to rest of site through open space system.

Road Layout: uses preferred E to orientation parallel to slopes to reduce grading requirements and channel breezes and cold winds.

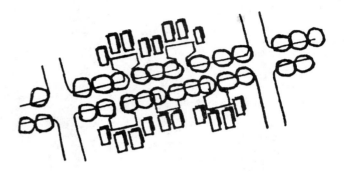

Building Relationships

Wind Response: orientation and buffering

Response to wind buffering in the temperate regions follows the same orientation guidelines based on the Mattingly and Peters study and identified for the cool region.

Sun Response: orientation and exposure

Sunlight probability for the temperate region of the country is in the mid range (.45 winter, 170 summer). While utilization of solar heating is feasible building relationships for all developments should be considered in terms of maximizing winter sun.

Orientation S 17.5° SE is optimum. Also in winter, south facing vertical walls receive twice as much sun as other orientations. If a building lot doesn't have this orientation, siting adjustments through set back flexibility can begin to approach it.

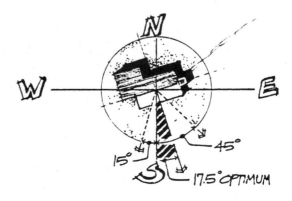

The following guidelines are recommended for optimizing winter sun.

Avoid locating buildings immediately to the north of large evergreen vegetation which would reduce solar radiation by blocking winter sun (ASLA p. 173).

Preferable building sites are those shaded by existing deciduous vegetation which does not block winter sun and provides summer shade.

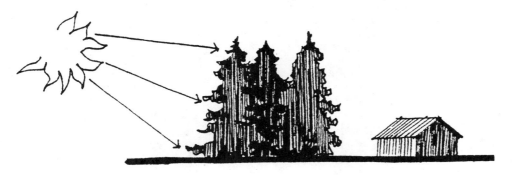

Cluster buildings in small villages to provide protected enclaves. Cluster buildings for heat absorption S-SE. Protect west and east exposures of buildings by "row house" pattern.

Generally linear plans are recommended in mild or balanced climate regions to permit the maximum use of natural radiant heat and through ventilation (BSIC p. 8).

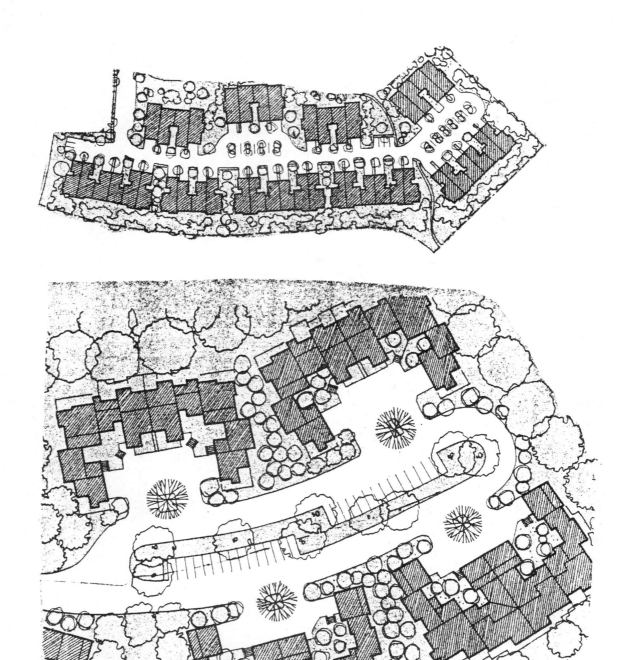

■Locate lightly used activity area on northern exposure.
■Locate exterior living space and active zones on S-SE.
■Locate lawns near structures.
■Locate overhands, S side on low structures.
■Locate egg crate sun shades on East and West exposure.
■Locate vertical fins, N side.

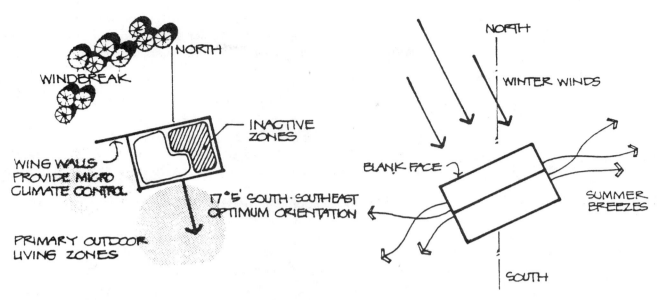

WINDBREAK

NORTH

INACTIVE ZONES

WING WALLS PROVIDE MICRO CLIMATE CONTROL

17°5' SOUTH-SOUTHEAST OPTIMUM ORIENTATION

PRIMARY OUTDOOR LIVING ZONES

NORTH

WINTER WINDS

BLANK FACE

SUMMER BREEZES

SOUTH

Combined Responses to Sun and Wind:

- ■Windbreaks against NW wind evergreen preferred.
- ■All vegetation: permit open S-SW breezes in summer.
- ■Deciduous: shade and open for S-SW breezes, best on E and W sides.
- ■Use roadways to channel and direct desirable breezes or direct unwanted cold winds.
- ■E-W street orientation is preferred.

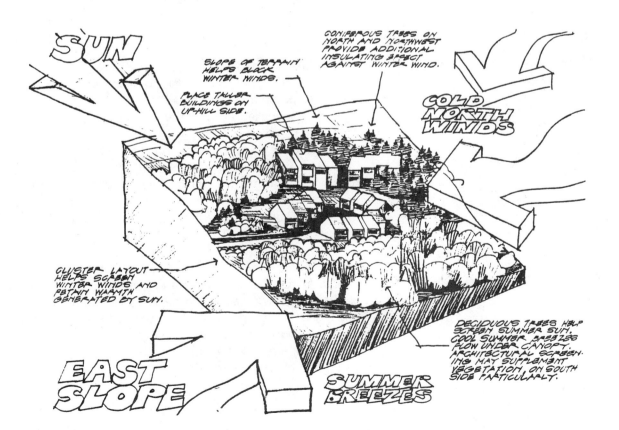

SUN

COLD NORTH WINDS

EAST SLOPE

SUMMER BREEZES

SLOPE OF TERRAIN HELPS BLOCK WINTER WINDS.

PLACE TALLER BUILDINGS ON UPHILL SIDE.

CONIFEROUS TREES ON NORTH AND NORTHWEST PROVIDE ADDITIONAL INSULATING EFFECT AGAINST WINTER WIND.

CLUSTER LAYOUT HELPS SCREEN WINTER WINDS AND RETAIN WARMTH GENERATED BY SUN.

DECIDUOUS TREES HELP SCREEN SUMMER SUN. COOL SUMMER BREEZES FLOW UNDER CANOPY. ARCHITECTURAL SCREENING MAY SUPPLEMENT VEGETATION ON SOUTH SIDE PARTICULARLY.

Additional guidelines identified during data collection are based on planning and design studies conducted by the firm of Rahenkamp, Sachs, Wells and Associates as follows (ASLA pp.203-213).

Mixed Residential Uses:
Pine Run, PUD in Camden, New Jersey, mixed residential development of single family, townhouses and garden apartments, was sited in response to a valley situation with typical cold air flow patterns at the valley bottom.

Development pattern essentially sited high density residential uses on steeper woodland areas and low density single family residential uses in flat open farm land. Multi-family units were designed as a modular unit with a standard angle to bend the building to the existing gradient and employ contour clustering. The density pattern and contour clustering system for multi-family housing produced greater protection from winter wind by capitalizing on the existing air flow pattern and vegetation. Clearing costs were reduced between $250-$270 per acre. Siting buildings within fifteen feet of existing vegetation reduced landscape costs between $300-$500 per unit (ASLA p. 204).

In terms of building unit relationships, experimentation in energy conservation in the design of single family units was employed as follows.

Single Family Units
- Townhouse type unit with an attached garage was placed for standing on a small lot.
- Fencing was used around the unit creating an internal housing space.
- Buffering of exterior and end walls with the system resulted in reduced winter heating requirements.
- Also minimum disturbance of existing topography resulted in savings between $.10 to $.15 per sq. ft. for bank stabilization plantings with an annual maintenance savings of $1,900 per year.

Multi-Family Units: concentrated on steeper wooded sites with contour clustering and a modular unit. A loop parking lot was created on one side of the building with open space and pedestrian connectors on the other side. This unit design and layout provides access to summer breezes while existing vegetation buffers the structures from the winter wind.

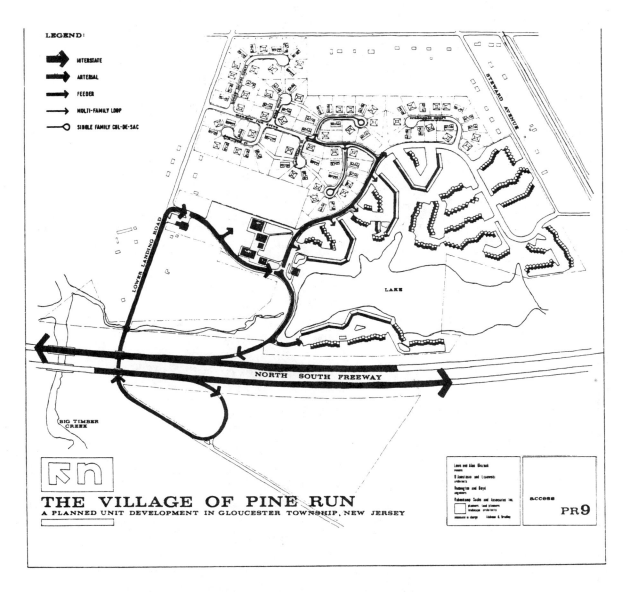

STEWARD AVENUE

LOWER LANDING ROAD

LAKE

NORTH SOUTH FREEWAY

BIG TIMBER CREEK

THE VILLAGE OF PINE RUN
A PLANNED UNIT DEVELOPMENT IN GLOUCESTER TOWNSHIP, NEW JERSEY

access

PR9

Development Suitability for various uses, structural types and use intensity.

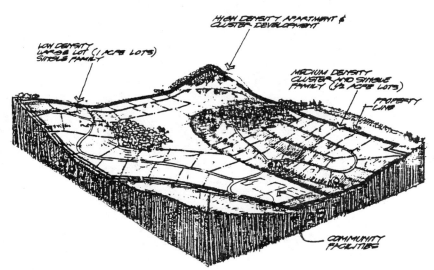

LOW DENSITY LARGE LOT (1 ACRE LOTS) SINGLE FAMILY

HIGH DENSITY APARTMENT & CLUSTER DEVELOPMENT

MEDIUM DENSITY CLUSTER AND SINGLE FAMILY (½ ACRE LOTS)

PROPERTY LINE

COMMUNITY FACILITIES

hot humid region

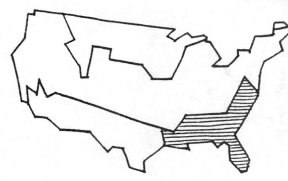

CLIMATE CHARACTERISTICS

Winter
- Sunlight probability .55 (BSIC p. 22)
- Maximum solar radiation on south vertical walls 4x intensity of E & W

Summer
- Sunlight probability .75 (BSIC p. 25)
- Maximum solar radiation E & W facing vertical walls 2-3x intensity of S

Year Round
- Typical warm climate with small yearly variation
- 3/4 of the year's temperature— 65°–85° range
- High humidity due to high precipitation, effects of ocean and underdrained lowland evaporation (ASLA p. 147)
- Shading required 75% of year (Olgyay p. 173)

DESIGN CRITERIA

Summer
- Cooling requirement—633 hours (BSIC p. 25)
- Reduce heat production and radiation gain
- Promote evaporation loss
- Minimize sun, maximize wind

Winter
- Heating requirements—2,000 degree days (BSIC p. 22)

GROSS SITE SELECTION GUIDELINES

Look for sites with slope elevation, orientation, vegetation and wind pattern to increase summer and winter cooling and decrease radiation effects. Look for wind patterns in relationship to:

Landform
- Mountainous or hilly coastal sites which are more open to breezes.
- Avoid coastal lowland areas if dampened by the effect of dense planting or buildings which act as wind breaks.

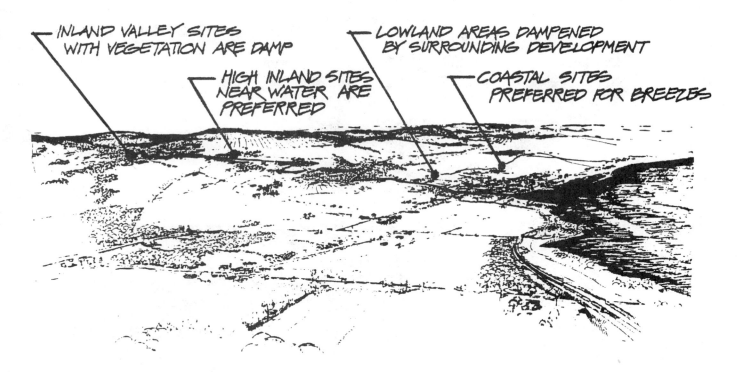

INLAND VALLEY SITES
WITH VEGETATION ARE DAMP

HIGH INLAND SITES
NEAR WATER ARE
PREFERRED

LOWLAND AREAS DAMPENED
BY SURROUNDING DEVELOPMENT

COASTAL SITES
PREFERRED FOR BREEZES

Water
■Look for site with water or near water. Inland water
bodies produce cooling effect when breezes are present
(ASLA p. 148).

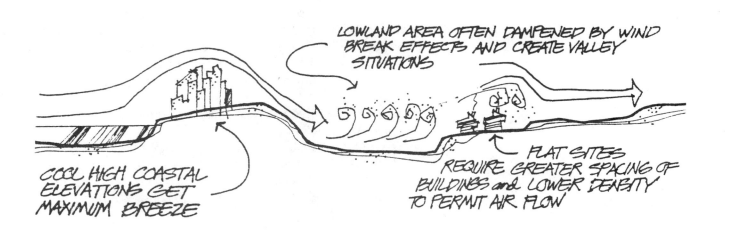

LOWLAND AREA OFTEN DAMPENED BY WIND
BREAK EFFECTS AND CREATE VALLEY
SITUATIONS

COOL HIGH COASTAL
ELEVATIONS GET
MAXIMUM BREEZE

FLAT SITES
REQUIRE GREATER SPACING OF
BUILDINGS and LOWER DENSITY
TO PERMIT AIR FLOW

107

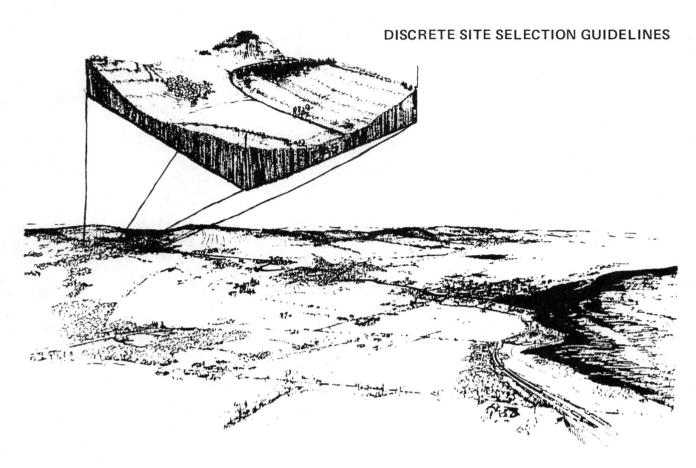

Site Analysis

Analysis and selection of residential sites should be based on the following general criteria:

- Identify and select sites with the greatest exposure to summer breezes for ventilation and cooling. Most important features are wind exposure, topography, and vegetation.
- On inland sites look for high elevations and slopes to capture breezes.
- Avoid low inland sites which are poorly drained or sites with dense adjacent development. These are potential air pockets.
- Look for wooded sites with high canopy trees which permit air movement and provide shade.

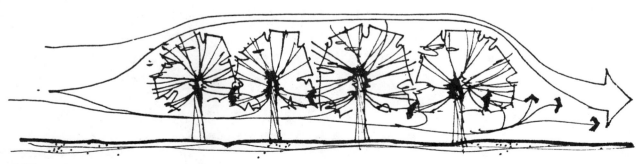

Avoid sites with dense low canopy trees which block breezes and trap humidity in dead air pockets (ASLA p. 148).

Seasonal Wind and Cold Air Flow
- Higher elevations preferred for ventilation or cooling.
- Avoid low spots subject to dampness, frost and potential dead air pockets.
- Seek sites protected from night time cold air flow but with exposure to breezes.

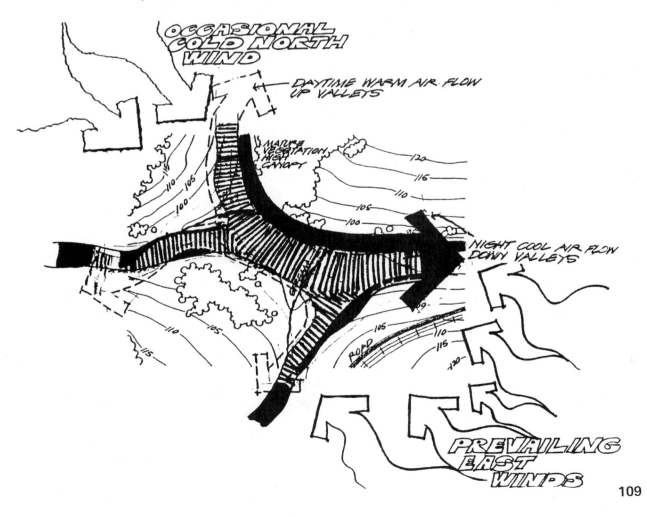

109

Slope Direction and Gradient
- Avoid slopes in excess of 20% due to erosion problems.
- Avoid high water table due to associated dampness and humidity.

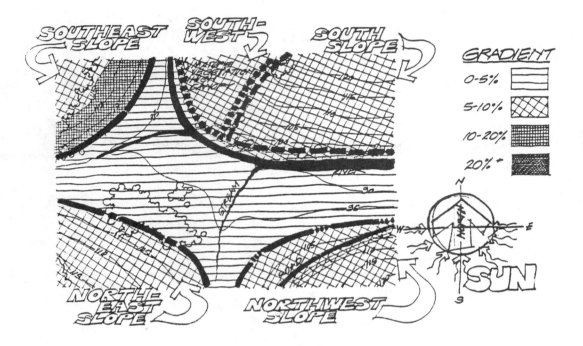

Site Selection

Site selection should be based on combined consideration of:

Winter and Summer Climate Influences related to existing land form, vegetation and natural features.

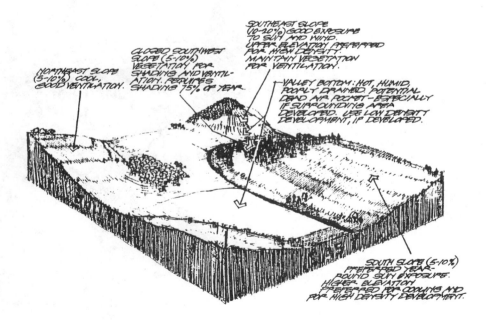

Recommended guidelines for various uses and structural types are as follows (Olgyay p. 173-177).

Residential Structures: Select southern and northern slopes which are cooler year round. Avoid east and west slopes which are too hot.
Tall structures: Select sites down wind of lower structures so as not to prevent through ventilation.
Exterior Open Space: Select shaded areas.

Plan Layout

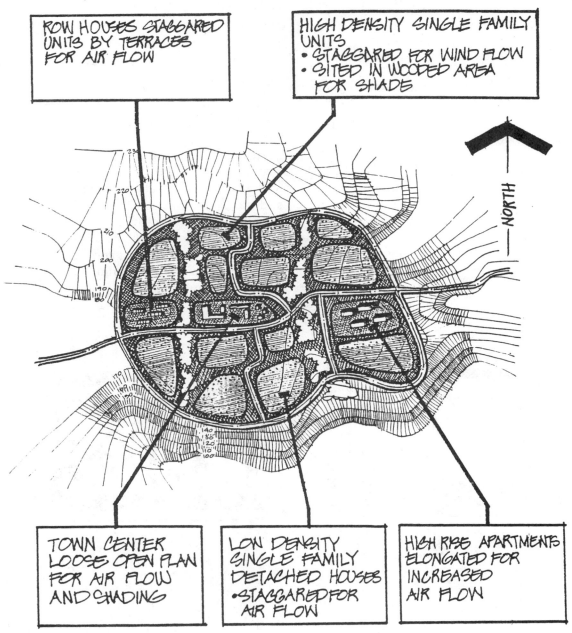

ROW HOUSES STAGGARED UNITS BY TERRACES FOR AIR FLOW

HIGH DENSITY SINGLE FAMILY UNITS
• STAGGARED FOR WIND FLOW
• SITED IN WOODED AREA FOR SHADE

NORTH

TOWN CENTER LOOSE OPEN PLAN FOR AIR FLOW AND SHADING

LOW DENSITY SINGLE FAMILY DETACHED HOUSES
• STAGGARED FOR AIR FLOW

HIGH RISE APARTMENTS ELONGATED FOR INCREASED AIR FLOW

Guidelines for plan layout are as follows (Olgyay p. 174-177).

- Loose scattered plans for air flow.
- Low building coverage with high density (apartments) or terraced sites.
- Higher buildings coverage with low density single family detached units on sloping sites. Separate structures encouraged to permit air flow.
- Town center—open loose connected structures for air movement and overall shading.
- Street layout—uses E-W pattern to channel air flow.

The generally desired character of development is low density layout with consistent sun wind orientation.

Building Relationships

Wind Response: orientation and layout

The goal in the hot humid region is to use the wind by slowing it down. In grouping buildings, remember that if wind encounters an inlet and an outlet in direct alignment with its direction, it will pass through the interviewing space in a narrowly defined path with minimal ventilation resulting.

To increase ventilation within a building cluster you want to cause change in direction between inlet and outlet and create an interior circulation air current to improve overall ventilation (Hastings and Crenshaw p. 1-32).

In siting detached structures or clustering attached structures encourage maximum ventilation by orienting structures and groupings as follows.

When overall orientation is askew to the direction of the prevailing wind, use building groupings with openings on opposite sides for maximum local air circulation and cooling (Hastings and Crenshaw pp. 1-31).

When overall orientation is perpendicular to the prevailing wind, group buildings with openings on adjacent sides for maximum local air circulation and cooling.

The following series illustrates building clustering responses to prevailing wind pattern.

The most undesirable building layout is one with major openings perpendicular to the prevailing wind. In this situation wind funnels through at maximum velocity with minimum cooling effect. Dead air pockets are created near units and minimum cooling results.

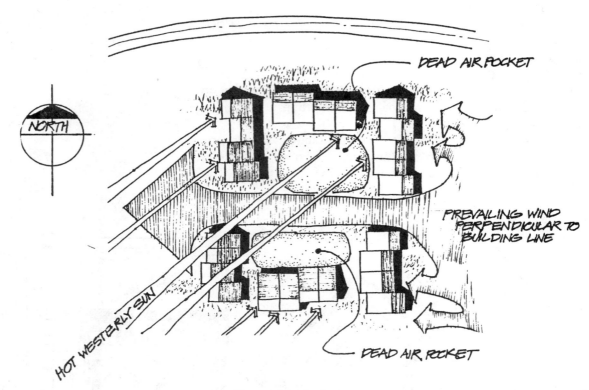

DEAD AIR POCKET

NORTH

PREVAILING WIND PERPENDICULAR TO BUILDING LINE

HOT WESTERLY SUN

DEAD AIR POCKET

For this wind pattern maximum response in terms of building grouping is to site cluster with major openings diagonal to the prevailing wind flow.

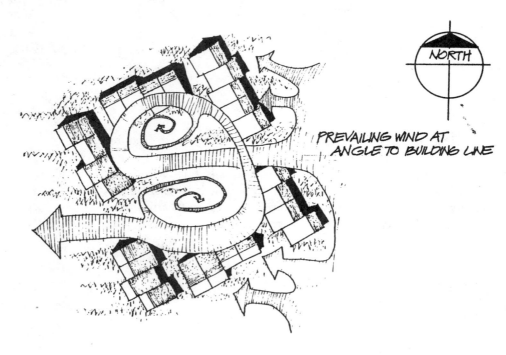

NORTH

PREVAILING WIND AT ANGLE TO BUILDING LINE

This clustering pattern results in maximum local ventilation air pockets and cools surrounding units.

When building clusters are perpendicular to the prevailing wind major openings in cluster groupings are desired as illustrated. This arrangement encourages a natural air flow for cooling and eliminates dead air pockets.

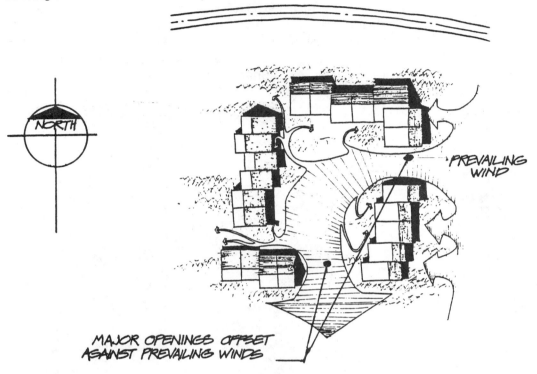

NORTH

PREVAILING WIND

MAJOR OPENINGS OFFSET AGAINST PREVAILING WINDS

In addition to clustering the units as previously illustrated further cooling can be induced by locating attached garages on westerly (hot afternoon sun) exposures.

This arrangement of garage units results in reduced radiation absorption into living areas and further provides a buffer from cold winter winds.

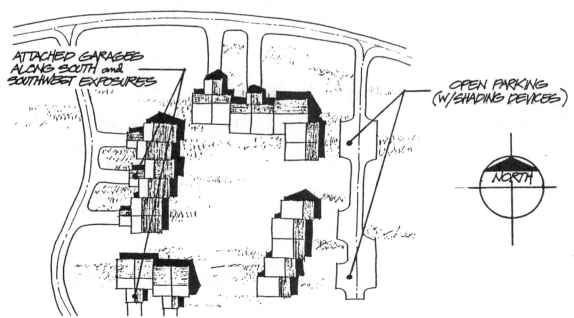

ATTACHED GARAGES ALONG SOUTH and SOUTHWEST EXPOSURES

OPEN PARKING (W/SHADING DEVICES)

NORTH

Finally, consideration should be given to placement and selection of plant materials such as mature high canopy deciduous vegetation planted on westerly and south westerly exposures.

This planting arrangement results in summer shading and evaporative cooling in association with the wind. It also increases local air circulation further by dispensing air flow patterns. Again consideration is given to buffering of cold winter winds and minimizing radiation on western exposure.

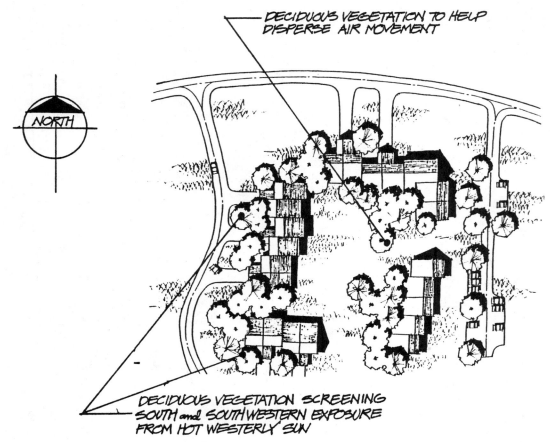

DECIDUOUS VEGETATION TO HELP DISPERSE AIR MOVEMENT

NORTH

DECIDUOUS VEGETATION SCREENING SOUTH and SOUTHWESTERN EXPOSURE FROM HOT WESTERLY SUN

DESIGN GUIDELINES

Sunlight probability for the hot, humid region of the country is higher than other regions except the hot arid (.55 winter, .75 summer) and should be considered in terms of minimizing overheating effects.

Orientation S 5° SE is optimum for residential developments. East and west slopes should be avoided and balance of sun and breezes is preferred.

The following guidelines are recommended for maximum summer shade and winter sun (ASLA p. 76).

■ Locate structure in deciduous woods or plant deciduous woods or plant deciduous trees for summer shade and winter sun.

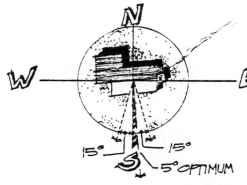

N

W

E

15°   15°

S

5° OPTIMUM

116

- In summer the structure should be shaded in the late morning and afternoon.
- Frequently used outdoor spaces should be shaded from late morning and afternoon sun in summer
- Pedestrian circulation routes should be shaded by vegetation, canopies, pergolas, or arcades.
- Deciduous trees should be located on the south side of pedestrian paths to allow winter sun.

Multi-Family Houses (Townhouses & Apartments)
- Make linear axis of structures E-W to reduce wall surface exposed to radiation.
- Cluster townhouses or apartments to catch breeze and use open space to channel wind.
- E-W linear axis of units for minimum sun exposure in the afternoon.
- Open spaces to channel breezes.

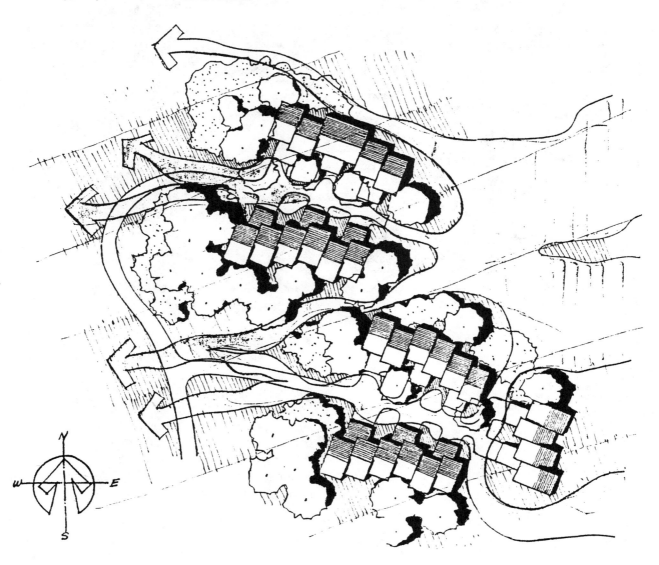

Single Family Housing

■Stagger set back lines of structures to permit air movement throughout.

■Utilize property lines in common as windbreaks to channel easterly breezes and shading for both properties (ASLA p. 166-167).

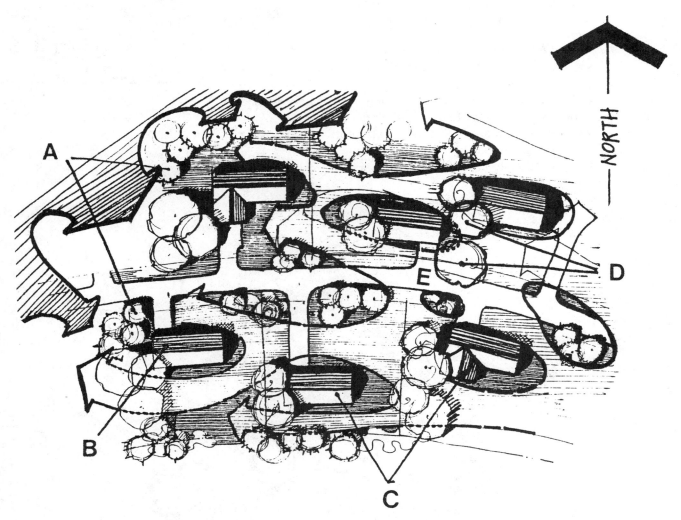

A. Dense evergreen plantings to buffer cold winter winds

B. Garage west end to block hot west sun

C. Setbacks varied to allow cooling, shading, ventilation

D. High canopy ventilation planted on southwest: shades and cools in summer, maximizes winter sun, minimizes dead air pockets

E. Major street on E-W maximizes air flow and provides optimum orientation

hot arid region

CLIMATE CHARACTERISTICS

Winter
- Sunlight probability .55 (BSIC p. 22)
- Maximum radiation on S facing vertical walls (4x E & W)

Summer
- Sunlight probability 0.75
- Maximum radiation E & W facing walls (2-3x S)
- Peak afternoon sun 3 p.m.

Year Round
- Clear skies, dry atmosphere, low precipitation
- Extended periods of overheating, large diurnal temperature range
- Wind generally along E-W axis with diurnal variations

DESIGN CRITERIA

Summer
- Cooling requirements 633 hrs (BSIC p. 23-24)
- Maximize shade
- Maximize air movement

Year Round
- Maximize humidity
- Reduce effects of daytime radiation
- Maximize morning and afternoon shade

GROSS SITE SELECTION GUIDELINES

Look for sites with slope orientation to *reduce* daytime radiation. Look for:

Landform
- Low hillside locations where cool air can be controlled.
- Avoid valleys and concave landforms which are too cold at night.
- Select high altitudes and positive landforms to minimize radiation on trap during the day and minimize cold air drainage at night.
- Optimum location is well upon a southeast slope with a mountain to the west.

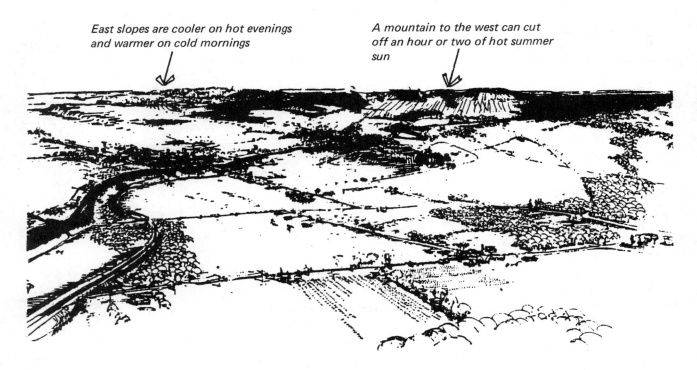

East slopes are cooler on hot evenings and warmer on cold mornings

A mountain to the west can cut off an hour or two of hot summer sun

Prevailing Winds
- East-west breezes are important for daily cooling.
- Avoid desert hot breezes.
- Maximize the humidity and cooling effects of breezes blown from across water.

DISCRETE SITE SELECTION GUIDELINES

Site Analysis

Analysis and selection of an optimum site should be based on the following general criteria:
- Identify and select sites with slope and orientation which reduce the effects of extreme daytime radiation and cooling requirements.
- Warm sites with S-SE orientations are preferred; avoid western orientation.

Summer Breezes and Winter Wind Pattern
- Select sites with preferred E-W breezes.
- Avoid sites subject to hot desert breezes.
- Cool air flow and evaporation associated with water locations is preferred.

Slope Direction and Gradient
- Avoid valley bottoms, especially narrow valleys.
- Select sites on higher locations which are cooler during the day and warmer at night.
- Select slopes with S-SE exposures.

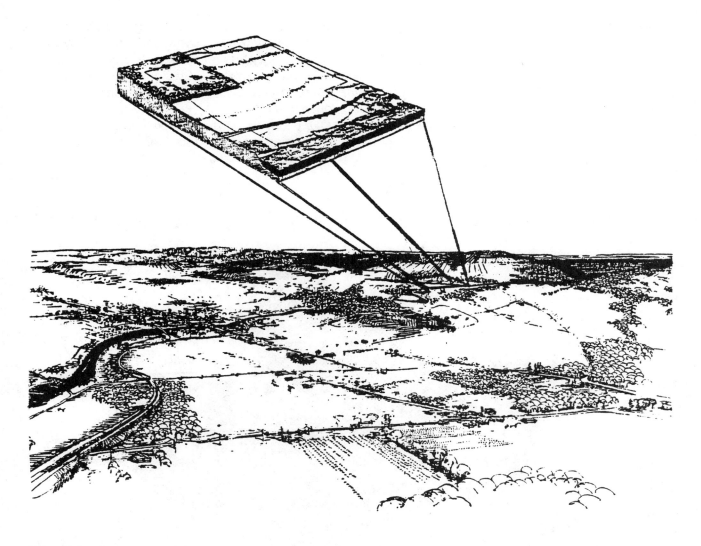

Site Selection

Site selection should be based on combined consideration of:

Winter and Summer Climate Influences:
- Select sites below middle of S-SE slope but above toe to benefit from cool air movements in early evening and warm air movements in early morning.
- Avoid valleys (ASLA p. 176).
- In valley situations, avoid valley floor and use cooler sites above for preferred evaporation.
- Ridge locations avoid heat traps and catch breezes.
- Select sites for subterranean or quasi-subterranean structures when possible, they are cooler and reduce energy requirements (ASLA p. 178).
- Select sites where existing vegetation or topography can be used to shade E & W walls.

Development Suitability for various uses and structural types. Studies by Olgyay recommend the following guidelines for various uses and structural types.
Residential Structures: Select S-SE slopes.
Exterior Open Spaces: Select half and full-shaded areas or public open space.
Exterior Living Spaces: Interior courtyards preferred, porches and patios generally too sunny, best location if used is on SE corner of house.

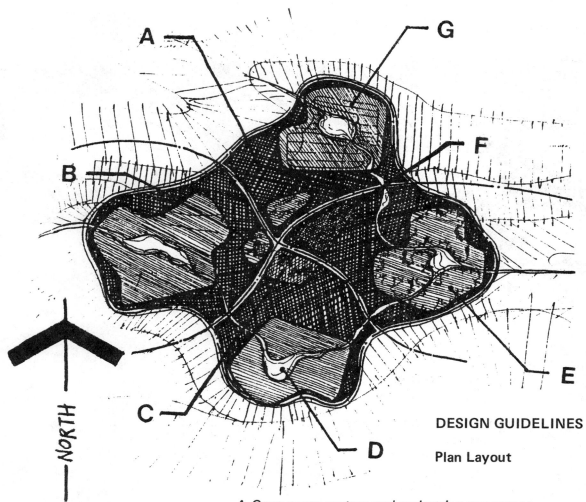

DESIGN GUIDELINES

Plan Layout

A. Open space system and pedestrian movement
B. School and recreation facilities
C. Court style housing permits interior shading and ventilation, maximum daytime cooling
D. Cluster housing near water body for maximum cooling
E. Locate clusters of high density single family housing to best utilize existing vegetation for shading
F. Town center commercial
G. High density single family detached, maximizes east, west streets for air movements

Guidelines for the Plan Layout are as follows (Olgyay p. 169-171).

High Density Single Family: Sited on south-southeast slopes utilizing compact patio arrangements with shared common walls and closed configurations where possible to maximize shading and evening radiation loss and cooling effect.

High Density Row House Units: Sited on south-southeast slopes with high building coverage and massing in cube configuration preferred to elongated linear attachments.

Exterior Living Space: Concentrated in interior courts formed by unit configuration utilizing building walls for shading.

Town Center: Centrally located, shaded and in walking distance of all units.

Street Layout: Narrow streets on east-west axis preferred to channel breezes. Streets should be shaded to reduce radiant heat.

Building Relationships
Sun Response: orientation and exposure

Sunlight probability for the hot, arid region is the highest in the country. Utilization of solar heating systems is extensively used. Whether using solar systems or conventional development, building relationships for all development should utilize maximum winter radiation.

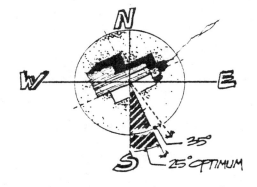

Optimum Orientation is 25° SE

The following guidelines are recommended for optimizing winter sun:
- Closed building arrangements around grass courts are preferred to promote heat loss through evaporation and cooling.
- Use compact building forms to expose minimum surface to radiation.
- Cluster houses to minimize building surface and utilize structures created by shade.

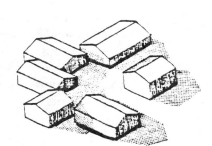

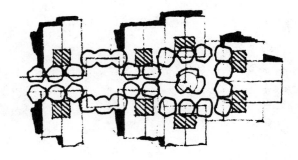

- Arrange dwellings around courtyard-like cooling wells for shading (ASLA p. 177).
- Provide close connections between residential and other uses.
- Small parking bays reduce sun pockets.
- Shaded and dense layout desirable.
  Interior courtyards provide shading opportunities.
- Locate activity areas to the southeast to collect early morning sun and permit afternoon shade.
- Orient outdoor living and working areas according to the times during which they are most frequently used, seeking cooler locations for afternoon activities and warmer locations for evening activities.
- Locate outdoor utility, storage and similar areas in those locations on the site not otherwise better used for living and working activities.
- Avoid west windows and bedrooms.
- Locate carport, garage, tool shed and blank walls on west to minimize sun impact.

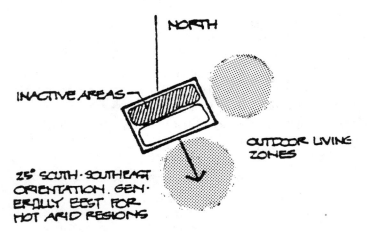

Solar planned communities in California which have been reviewed (Village Homes, Inc., City of Davis, Tandem, Indio) have identified the following guidelines for maximum solar utilization and energy conserving design.

Lot and Street Orientation

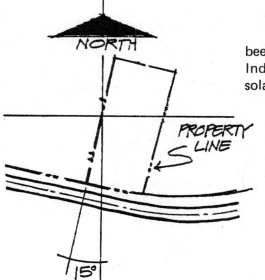

Recommended lot orientation is within 15° of south on the long axis of the lot to minimize southern sun for heating. Recommended street orientation is E.W. to channel breezes and permit maximum number of south oriented lots.

The lot and street layout illustrated demonstrates the feasibility of optimum southern lot orientation in a typical subdivision. By redesign of street layout to a major E.W. pattern (example shown) most residential lots in a traditional subdivision are given an orientation within 15° of south (Hammond April 1977).

The above Stanley Davis subdivision was redesigned from east-west to north-south orientation

A similar example of this type is the Tandem development in California where again E.W. orientation of residential feeder streets permits maximum number of south oriented lots (Tandem, unpublished).

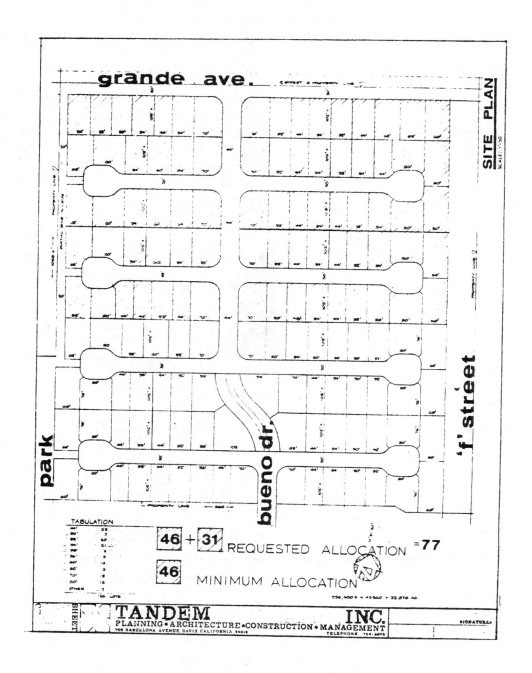

Alternative Plan Layout

In examining traditional subdivision layouts, communities surveyed have identified wasted space and unused development costs for traditional linear subdivision blocks as illustrated in the following plan (Hammond 1977 p. 38).

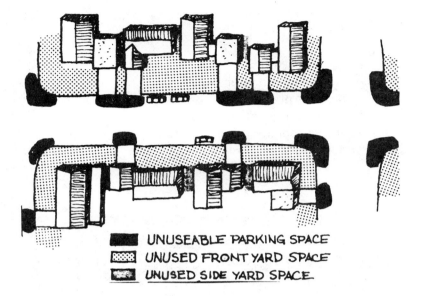

UNUSEABLE PARKING SPACE

UNUSED FRONT YARD SPACE

UNUSED SIDE YARD SPACE

Recommended alternative development patterns include:
- Compact arrangement around common court permits maximum daytime shading and evening cooling.
- Land developed on zero foot lot line requires conditional use permit and set back flexibility but permits common wall construction which reduces undesired heat loss and gain and construction costs (Hammond 1977 p. 39).

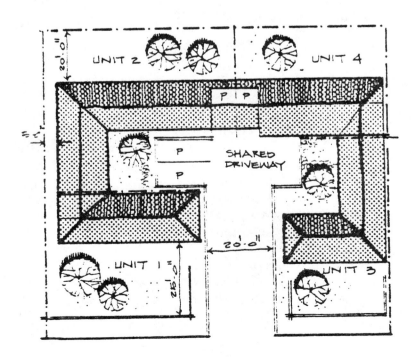

- Reduced paving for driveway and court system minimizes reflective heat surfaces, impervious drainage and facilitates shading.
- System suited for adaptation to passive or active solar heating systems.

Back-to-Back 4-Unit Cluster
- Alternative to compact courtyard where greater privacy is provided.
- Adapts to E.W. street system.
- Permits common wall construction which reduces undesired heat loss and heat gain and construction costs (Hammond 1977 pp. 38-39).

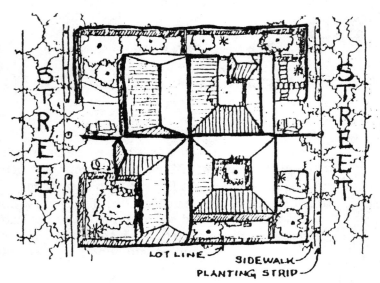

BACK TO BACK 4-UNIT CLUSTER

Increased Setback Flexibility
When single family lots and detached housing are mandated by program, client or zoning restrictions, or when lot orientation deviates from optimum, preferred unit orientation can be accomplished through flexible setback arrangements as illustrated.
- Permits full exposure to southern winter sun for heating.
- Prevents shading where undesirable.
- Permits southern orientation of structure where lot orientation is not N-S (Hammond 1977 p. 39).

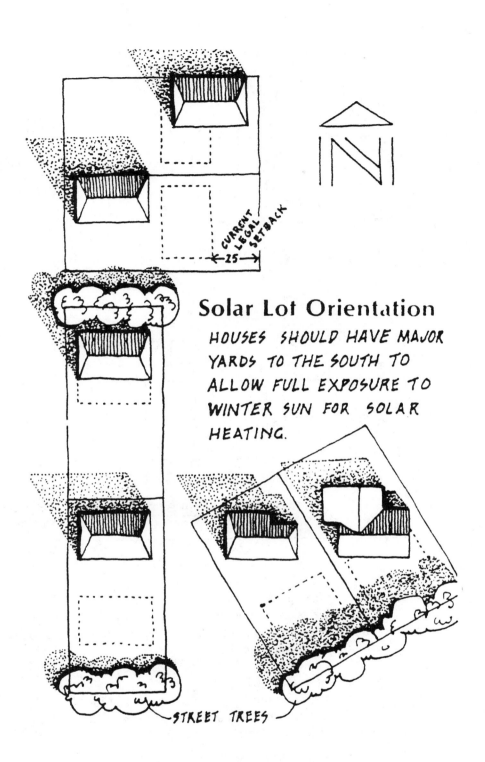

Solar Lot Orientation

HOUSES SHOULD HAVE MAJOR YARDS TO THE SOUTH TO ALLOW FULL EXPOSURE TO WINTER SUN FOR SOLAR HEATING.

CURRENT LEGAL SETBACK

25

N

STREET TREES

Flexible setbacks on detached dwelling units also permit optimum solar access on both sides of the street as illustrated.

10' rear yard setbacks on north side of the street provide large front yard living space on south, permitting developer flexibility and varied visual pattern by reversing building plan. This is dependent on location from front to back (Tandem, unpublished).

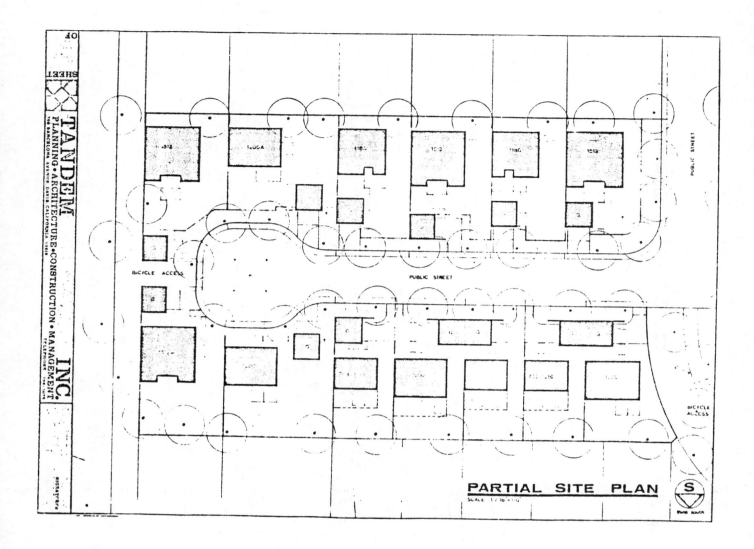

PARTIAL SITE PLAN
SCALE 1/16"=1'0"

Measurement of Energy Surveys

Communities examined identified the following energy savings for combined solar planning of new communities and retrofit of existing development.

Davis (Hammond 1977 p. 1)
Total energy savings—50% (Hammond 1977 p.1)
Energy savings for peak summer cooling—10%
Total growth in electricity use—0

Tandem Properties (Tandem, unpublished)
Total solar heating efficiency—90%
Total cooling efficiency—100%

Reduced Street Width
Several California energy conscious developments (Davis, Mission Viejo, Tandem, Village Homes) have advocated reduced street widths for the following reasons:
■ Surface temperature of asphalt on a 90° day can reach 140°

citations

AIA Research Corporation. **Energy Conservation in Building Design**. Washington, D.C.: August 1976, pp. 61, 65.

AIA Research Corporation. **Solar Dwelling Design Concepts**. U.S. Dept. of Housing & Urban Development, Office of Policy Development & Research: May 1976, pp. 64-73.

American Society of Landscape Architects Foundation. **Landscape Planning for Energy Conservation**. Reston, Virginia: Environmental Design Press. 1977.

Bainbridge, David A., & Hammond, Jonathan. **Planning for Energy Conservation**: Prepared for the City of Davis, California June 1, 1976. Funded by HUD Innovative Project Grant B-75-31-06-001. Living Systems, Winters, California.

BSIC/EFL. **BSIC/EFL Energy Workbook: Energy Conservation and the Building Shell**. Menlo Park, California: Building Systems Information Clearing House Educational Facilities Laboratories, Inc. July 1974.

Davis, Albert J. & Schubert, Robert P. **Alternative Natural Energy Sources in Building Design**. Van Nostrand Reinhold Company. 1977.

Fitch, James Marston. "This House Does the Impossible." **House Beautiful**. March 1951.

Geiger, Rudolf. **The Climate Near the Ground**. Cambridge, Massachusetts: Harvard University Press. 1957.

Goldberg, Philip. "Planning with Energy." Philadelphia, Pennsylvania: Rahenkamp, Sachs, Wells and Associates, Inc. January 1975.

Hammond, Jonathan et al., **A Strategy for Energy Conservation**: Proposed energy conservation and solar utilization ordinance for the City of Davis, California. Living Systems, Winters, California. 1974.

Hammond, Jonathan, et al. **Davis Energy Conservation Report**: Practical use of the sun. Living Systems, Winters, California. HUD Innovative Project B-75-51-05-001. April 1977.

Hammond, Jonathan, & Hart, M. **The Davis Energy Conservation Code**. Living Systems.

Harwood, Corbin Crews. **Using Land to Save Energy**. Cambridge, Massachusetts: Ballinger Publishing Company. c. 1977.

Hastings, S. Robert & Crenshaw, Richard W. **Window Design Strategies to Conserve Energy**. Washington, D.C.: Architectural Research Section, Center for Building Technology, Institute for Applied Technology, National Bureau of Standards. June 1977.

Howland, Joseph E. "How Privacy Can Increase Your Living Space and Improve Your Climate." **House Beautiful**. June 1950, pp. 101, 102.

Kaminsky, Jacob. "Environmental Characteristics Planning: An Alternative Approach to Physical Planning." Baltimore, Maryland: Regional Planning Council. July 1972.

Kroner, Walter M. & Haviland, David. **Passive Energy Technologies for Residential Construction**. Center for Architectural Research, Rensselaer Polytechnic Institute.

Langeweische, Wolfgang. "Can You Control the Wind?" **House Beautiful**. June 1950, pp. 88, 91, 97, 204.

Langeweische, Wolfgang. "How to Control the Sun." **House Beautiful**. March 1950, p. 131.

Langeweische, Wolfgang. "How to Fix Your Private Climate." **House Beautiful**. October 1949, pp. 108, 147, 150, 152, 192.

Langeweische, Wolfgang. "How to Live Comfortably in the Southwest Desert." **House Beautiful**. April 1950, pp. 205-207.

Langeweische, Wolfgang. "How to Manipulate Sun and Shade." **House Beautiful**. July 1950.

Langeweische, Wolfgang. "How to Pick Your Private Climate." **House Beautiful**. October 1949, pp. 147, 148.

Langeweische, Wolfgang. "Wind Control in Hot Weather." **House Beautiful**. June 1950, pp. 124, 165.

Leckie, Jim; Master, Gil; White House, Harry & Young, Lily. **Other Homes and Garbage: Designs for Self-Sufficient Living**. Sierra Club Books. 1975.

Lynch, Kevin. **Site Planning**, 2nd edition. Cambridge, Massachusetts: The M.I.T. Press c. 1971.

Mattingly, George E., & Peters, Eugene F. **Wind and Trees—Air Infiltration Effects on Energy in Housing**. Center for Environmental Studies Report No. 20. Princeton, New Jersey. May 1975.

NAHB. **The Builder's Guide to Energy Conservation**. 1974.

National Science Foundation Research Applications Directorate. **The Use of Earth Covered Buildings**: Alternatives in energy conservation. Proceedings of a conference held in Fort Worth, Texas. July 9-12, 1975.

Olgyay, Victor. **Design with Climate**. Princeton Univeristy Press, Princeton, New Jersey. 1963.

Read, Ralph A. "Tree Windbreaks for the Central Plains." U.S. Forest Service Agriculture Handbook, No. 250. February 1964.

Reimann-Buechner-Crandall Partnership. **Site Specific Energy Consumption Patterns**. Joint study with the Niagara-Mohawk Power Corporation of Central New York. 1977. Unpublished.

Siple, Paul E. "15,750,000 Americans Live in This Climate." **House Beautiful**. November 1949.

Smay, V. Elaine."Underground Houses." **Popular Science**. April 1977, pp. 84-89, 155.

Tandem Properties, Inc. **Tandem**. Barcelona Avenue, Davis, California. (Unpublished).

Wright, Henry. "How to Put a Harness on the Sun." **House Beautiful**. October 1949, pp. 153, 158, 185.

bibliography

Abt Associates, Inc., **In the Bank or Up the Chimney,** Prepared for the Office of Policy Development and Research, Division of Energy, Building Technology and Standards, U.S. Department of Housing and Urban Development, Washington, D.C., 1977, 73 pp.

AIA Research Corporation. **Energy Conservation in Building Design.** Washington, D.C.: August 1976, pp. 61, 65.

AIA Research Corporation. **New Design Concepts for Energy Conserving Buildings.** Washington, D.C.: 1976.

AIA Research Corporation. **Solar Dwelling Design Concepts.** U.S. Dept. of Housing & Urban Development, Office of Policy Development & Research: May 1976, pp. 64-73.

Abelson, Philip H. **Energy: Use, Conservation & Supply: A Special Science Compendium.** Washington, D.C.: American Association for Advancement of Science. 1974.

Agle, Charles K. **The Energy Crisis and Community Planning.** New Jersey Federation of Planning Officials. 1974.

Allen, Booz. **Interaction of Land Use Patterns and Residential Energy Conservation.** Bethesda, Maryland: Hamilton, Inc. 1976.

Alternative Sources of Energy. Issue No. 26; Special Shelter Section, Part II. June 1977.

American Institute of Architects. **AIA Energy Notebook, Information Service on Energy and the Built Environment.**

American Institute of Planners. **Put Energy in Your Planning—A How-To Guide for Community Planners.** June 1976.

American Society of Landscape Architects. **Landscape Planning for Energy Conservation.** Reston, Virginia: Environmental Design Press. 1977.

American Society of Planning Officials. **Energy Conservation: What Planners Can Do About It.**

Anderson, Bruce. **The Solar Home Book: Heating, Cooling & Designing with the Sun.** Cheshire Books. 1976.

Argonne National Laboratory, Energy and Environmental Systems Division, **Site and Neighborhood Design for Energy Conservation: 5 Case Studies,** U.S. Department of Energy, Washington, D.C., 1982, 581 pp.

Aronin, Jeffrey Elles. **Climate and Architecture.** New York: Reinhold Publications. 1953.

BSIC/EFL. **BSIC/EFL Energy Workbook: Energy Conservation and the Building Shield.** Menlo Park, California: Building Systems Information Clearing House Educational Facilities Laboratories, Inc. July 1974.

Bainbridge, David A. & Hammond, Jonathan. **Planning for Energy Conservation:** Prepared for the City of Davis, California. Winters, California: Living Systems. June 1, 1976.

Bainbridge, David & Moore, M. **Bikeway Planning and Design: A Primer**. Living Systems. 1979.

Berman, Sand & Silverstein, S. D. **Energy Conservation & Window Systems, in Efficient Use of Energy**. The APS Studies on the Technical Aspects of the More Efficient Use of Energy. American Physical Society. 1975.

Brown, Theodore L. **Energy and the Environment**. Columbus, Ohio: Merrill, 1971.

Bullard, Clark W. **Energy and the Regional Planner**. University of Illinois at Urbana Center for Advanced Computation. May 1974.

Caborn, J. M. **Shelterbelts & Windbreaks**. London, Faber & Faber, c. 1965.

Calderson, C. M. & Manchauna, D. M. **Energy and Local Governments**. Arlington, Texas: University of Texas Institute of Urban Studies. September 1974.

Carroll, et al. **Land Use and Energy Utilization**. SUNY, Upton, New York; Brookhaven National Labs. November 1975.

Carroll, T. Owen; Nathans, Robert; Palmedo, P. F.; Stern, R. **The Planner's Energy Workbook: A User's Manual for Use and Energy Utilization Relationships**. Upton, New York: National Center for Analyzing Energy Systems, Brookhaven National Labs. Stoney Brook, New York: Institute for Energy Research, State of New York. October 1976.

Carrier, Roger. **Energy Conservation Through Urban Transportation Planning**. Harrisburg, Pennsylvania: Pennsylvania Transportation Institute. May 1975.

Caudill, W.W., Sherman E. Crites, Elmer G. Smith, **Some Genral Considerations in the Natural Ventilation of Buildings,** Texas A & M Engineering Experiement Station, College Station, Texas, 1951.

Chalmers, Bruce. **Energy**. New York: Academic Press. 1963.

Claremont: Center for California Public Affairs. **Energy: A Guide to Organizational & Informational Resources in U.S.** 1974.

Clark, Wilson. **Energy For Survival: The Alternative to Extinction**. Garden City, New York: Anchon Press. 1975.

Collins, Eugene N. **Conservation of Energy in Chattanooga: A Study of One City's Attempt to Legislatively Reduce Energy Consumption**. January 1975.

Columbia University; Taussig, R. T. **Commercial Energy Use**. Pergamon Press, Inc. 1976.

Conklin and Rossant, et al. **Reading the Energy Meter on Residential Development**. New York, New York: November 1976.

Cotruvo, Thomas D. **Energy Conservation & Urban Planning Implications for Land Use Patterns**. Philadelphia, Pennsylvania: University of Pennsylvania, Department of Civil and Urban Engineering. May 1976.

Davis, Albert J. & Schubert, Robert P. **Alternative Natural Energy Sources in Building Design**. Van Nostrand Reinhold Company.

Dawson, Joe. **Buying Solar**. Washington, D.C.: U.S. Government Printing Office (FEA/G-76/154). June 1976.

Deering, Robert B. **Technology of the Cooling Effect of Trees & Shrubs, Housing and Building in Hot Humid and Hot Dry Climates**. Davis, California: University of California.

Department of Water Resources. **Save Water**. Sacramento, California: 1976.

Dietz, Albert & Heney, George. **Dwelling House Construction**. Cambridge, Massachusetts: MIT Press. 1971.

Dober, Richard P. **Environmental Design**. "Climate & Regional Design." New York: Van Nostrand Reinhold. 1969.

Dole, Stephen H. **Energy Use and Conservation in the Residential Sector**. Santa Monica, California: Rand Corporation. June 1975.

Drew, J. & Fry, M. **Tropical Architecture in the Dry and Humid Zones**. New York: Reinhold Publications. 1964.

Editors of Sunset Books, **Sunset Homeowner's Guide to Solar Heating,** Lane Publishing Co., Menlo Park, CA. 94025, 1978, 96 pp.

Egler, Frank E. **Energy Conservation on the Home Grounds: The Role of Naturalistic Landscaping**, "The Home Lawn: Botanical Absurdity," edited by William A. Niering & Richard G. Goodwin. New London, Connecticut: The Connecticut Arboretum at Connecticut College. 1975.

Eisenhard, Robert M. **Building Energy Authority and Regulations Survey: State Activity**. Washington, D.C.: Office of Building Standards and Codes Services, Center for Building Technology, Institute for Applied Technology, National Bureau of Standards. March 1976.

Eldridge, Frank R. **Wind Machine**. Washington, D.C.: National Science Foundation. October 1975.

Energy Research and Development Administration. **An Analysis of Solar, Water, and Space Heating**. Washington, D.C.: U.S. Government Printing Office. November 1976.

EPA. **Bicycle Transportation**. Superintendent of Government Documents, Government Printing Office, Washington, D.C.: (GPO-055-000-00135-9). December 1974.

Erly, Duncan and Martin Jaffe, **Protecting Solar Access for Residential Developments: A Guidebook for Planning Officials,** U.S. Department of Housing and Urban Development, Washington, D.C., 1979, 154 pp.

Erly, Duncan and Martin Jaffe, **Site Planning for Solar Access: A Guidebook for Residential Developers and Site Planners,** U.S. Department of Housing and Urban Development, Washington, D.C., 1979, 149 pp.

Federal Energy Office. **Sensitivity of the Leisure Recreation Industry to the Energy Crisis**. Booz Allen & Hamilton, Inc. January 18, 1974.

Florida, University of. **Energy: The Power of the States**. Gainesville, Florida: Holland Law Center, Center for Governmental Responsibility. Jon L. Mills, et al. October 1975.

Fowler, John M. **Energy and the Environment**. New York: McGraw-Hill. 1975.

Friend, Gil & Morris, David. **Kilowatt Counter**. Alternative Sources of Energy Magazine. 1975.

Fry, Maxwell and Jane Drew, **Tropical Architecture in the Dry and Humid Zones,** Reinhold Publishing Co., New York, 1964.

Fry, M. & Drew J. **Tropical Architecture in the Humid Zone**. New York: Reinhold Publications. 1956.

Geiger, Rudolf. **The Climate Near the Ground**. Cambridge, Massachusetts: Harvard University Press. 1957.

Givonio, B. **Man, Climate and Architecture**. New York: Elsevier Publishing Company., 1969.

Goodwin, Richard H., et al. **Energy Conservation on the Home Grounds: The Role of Naturalistic Landscaping**, "Case Histories," edited by William A. Niering & Richard H. Goodwin. New London, Connecticut: The Connecticut Arboretum at Connecticut College. 1975.

Greenbrier Associates, **Case Study of Project Planning and Design for Energy Conservation - Greenbrier, Chesapeake, Virginia,** Final Report, Greenbrier Associates, Chesapeake, Virginia, 1980, Available from National Technical Information Service, U.S. Department of Commerce.

Greenbrier Associates, **Thoughtful Landscaping Saves Energy,** Pamphlet No. 1, Greenbrier Associates, Chesapeake, Virginia, 1977, 4 pp.

Greenbrier Associates, **Energy Saving Ideas for House Planning,** Pamphlet No. 2, Greenbrier Associates, Chesapeake, Virginia, 1977, 4 pp.

Greenbrier Associates, **Building Details That Save Energy,** Pamphlet No. 3 Greenbrier Associates, Chesapeake, Virginia, 1977, 4 pp.

Greenbrier Associates, **Why Double Glazing is Cheaper Than Single Glazing,** Pamphlet No. 4, Greenbrier Associates, Chesapeake, Virginia, 1977, 4 pp.

Greenbrier Associates, **A Greenhouse Can Save Energy,** Pamphlet No. 5 Greenbrier Associates, Chesapeake, Virginia, 1977, 4 pp.

Griffin, Charles W. **Energy Conservation in Buildings: Techniques for Economical Design**. Washington, D.C.: Construction Specifications Institute. 1974.

Hammond, J.M., Hunt R. Cramer and L. Neubauer, **A Strategy for Energy Conservation,** Energy Conservation Ordinance Project, City of Davis Davis, CA., 1974.

Harwood, Corbin Crews. **Using Land to Save Energy**. Cambridge, Massachusetts: Ballinger Publishing Company. c. 1977.

Hastings, S. Robert & Crenshaw, Richard W. **Window Design Strategies to Conserve Energy**. Washington, D.C.: Architectural Research Section, Center for Building Technology, Institute for Applied Technology, National Bureau of Standards. June 1977.

Hill, James E. & Richtmyer, Thomas E. **Retrofitting a Residence for Solar Heating & Cooling: The Design & Construction of the System**. Washington, D.C.: U.S. Department of Commerce, National Bureau of Standards. November 1975.

Hoffman Associates, Inc. **Comprehensive Community Planning for Energy Management and Construction**. April 1977.

Houghton, Edward Lewis & Carruthers, N. B. **Wind Forces on Buildings and Structures: An Introduction**. New York: John Wiley & Sons. 1976.

Housing and Home Finance Agency. **Application of Climatic Data to House Design**. Washington, D.C.: 1954.

Huldah, Carl. **Winona, Towards an Energy Conserving Community**. University of Minnesota. 1975.

Intergovernmental Regional and Special Programs, Office of. **Directory of Federal Agencies Engaged in Energy Related Activities**. Washington, D.C.: October 1975.

Institute of Rational Design. **Design and Control of Land Development in Suburban Communities**. 1973.

Jaffe, Martin and Duncan Erly, **Residential Solar Design Review,** A Manual on Community Architectural Controls and Solar Energy Use, Office of Policy Development and Research, U.S. Department of Housing and Urban Development, Washington, D.C., 1980, 86 pp.

Jones, J. W. & Hendrix, B. J. **Residential Energy Requirements & Opportunities for Energy Conservation**. Austin, Texas: Center for Energy Studies, University of Texas. 1975.

Knelman, Fred H. **Energy Conservation**. Ottawa, Canada: Science Council of Canada. 1975.

Knowles, Ralph L. **Energy and Form. An Ecological Approach to Urban Growth**. 1974.

Kroner, Walter M. & Haviland, David. **Passive Energy Technologies for Residential Construction**. Center for Architectural Research, Rensselaer Polytechnic Institute.

Kusuda, T. **National Bureau of Standards Load Determination Program for Heating and Cooling Loads in Buildings**. NBSIR 74-574. November 1974.

Land Design/Research, Inc. **Energy Conscious Development, Options for Land Use and Site Planning Regulations,** Fairfax County, Virginia, Fairfax County Board of Supervisors, 1981, Fairfax, Virginia, 28 pp.

Land Design/Research, Inc.,**Case Studies of Project Planning and Design for Energy Conservation(Burke Centre),** prepared for the U.S. Department of Energy, Community Systems Division, Washington, D.C. 1979.

Land Design/Research, Inc.**Planning for Housing - Development Alternatives for Better Environments,** Prepared for the National Association of Homebuilders, Washington, D.C., 1980.

Lang, Reg and Audrey Armour, **Planning Land to Conserve Energy,** Environment Canada, Land Use in Canada No. 25, Ottawa, 1982, 212 pp.

Large, David B. **Homes in Hidden Wastes: Potentials for Energy Conservation** (Washington: Conservation Foundation) in **Readings on Energy Conservation: Selected Materials**. Compiled by Congressional Research Service. Washington, D.C.: U.S. Government Printing Office. 1975.

Leckie, Jim; Masters, Gil; White House, Harry & Young, Lily. **Other Homes and Garbage: Designs for Self-Sufficient Living**. Sierra Club Books. 1975.

Lenihan, John. **Energy Resources & The Environment**. New York: Academic Press. 1976.

Little, Arthur D., Inc. **Residential and Commercial Energy Patterns 1970-1990**. Cambridge, Massachusetts. November 1974.

Lynch, Kevin. **Site Planning**, 2nd edition. Cambridge University Press. 1972.

Marier, Don & Abby. **Alternative Sources of Energy**. Volume No. 25. April 1977.

Martin, L. & March, L. **Urban Space & Structures**. Cambridge University Press. 1972.

Markham, S. F. **Climate and the Energy of Nations**. London: 1947.

McCullough, Sandra; Harnan, Mary S. & Gillan, Kim J. **A Handbook of Consideration for the Environmental Review Process**. Prepared for Southern Tier East Regional Planning Development Board and Broome County Environmental Management Council. June 1977.

Moffat, Ann Simons and Marc Schiler, **Landscape Design that Saves Energy**, William Morrow and Co., New York, 1981, 223 pp.

Moholy, N. S. **Native Genius in Anonymous Architecture**. New York: Horizon Press. 1957.

Morgan, M. Granger (edited by). **Energy & Man: Technical & Social Aspects of Energy**. New York: IEEE Press. c. 1975.

Morrison, Denton E. **Energy. A Bibliography of Soc-Sci & Related Literature**. New York: Garland Publishers. 1975.

NAHB. **The Builder's Guide to Energy Conservation**. 1974.

NSF/NASA Solar Energy Panel. **An Assessment of Solar Energy as a National Energy Resource**. Washington, D.C.: Government Printing Office. December 1972.

National Academy of the Sciences. **Optimizing the Use of Materials and Energy in Transportation and Construction**. Transportation Research Board. Special Report No. 166.

National Energy Information Center. **Directory of State Government Energy Related Agencies**. September 1975.

National League of Cities. **Energy Conservation in Buildings: New Role of Naturalistic Landscaping**. "Foreward," edited by William A. Niering & Richard H. Goodwin. New London, Connecticut: The Connecticut Arboretum at Connecticut College. 1975.

Odum, Howard T. & Elizabeth C. **Energy Basis for Man and Nature**. New York: McGraw Hill. 1976.

OECD Publication Center. **Better Towns with Less Traffic**. Washington, D.C.: OECD Conference Proceedings, Paris. April 14-16, 1975.

Olgyay, Victor. **Design with Climate**. Princeton, New Jersey: Princeton University Press. 1963.

Ontario Ministry of Housing, Project Planning Branch, **Residential Site Design and Energy Conservation,** Ontario Government Publication Service, 880 Bay St., 5th Floor, Toronto, Ontario, Canada, M7A, 1N8, 1981.

Ontario Ministry of Housing, **Saving Energy by Way of Site Design,** Ontario Government Publication Service, 880 Bay St., 5th Floor, Toronto, Ontario, Canada, M7A 1N8, 1980.

ORNL Review. **Ecology's 20-Year Space Program**. Oakridge, Tennessee: Winter 1975.

Pauker, Guy I. **Can Land Management Reduce Energy Consumption for Transportation**. Rand Corporation. May 1974.

Peski, Carolyn. **Solar Directory**. Ann Arbor, Michigan: Ann Arbor Science Publishers, Inc. 1975.

Peterson, Stephen R. **Building Science Series 64. Retrofitting Existing Housing for Energy Conservation: An Economic Analysis**. Washington, D.C.: U.S. Department of Commerce/National Bureau of Standards in Cooperation with Federal Energy Administration/ Energy Conservation and Environment. December 1974.

Portola Institute. **Energy Primer; Solar, Water, Wind & Biogradable Fuels**. Menlo Park, California. 1974.

Public Building Service of the General Services Administration, The. **Energy Conservation Guidelines for Existing Office Buildings**. February 1975.

Real Estate Research Corporation. **The Cost of Sprawl**. Washington, D.C.: U.S. Government Printing Office. April 1974.

Reimann-Buechner-Crandall Partnership, **Options for Passive Energy Conservation in Site Design,** prepared in cooperation with the Center for Landscape Architectural Education and Research, U.S. Department of Energy, Division of Buildings and Community Systems, 1978, 324 pp., Available from the National Techical Information Service(N.T.I.S.) 5285 Port Royal Rd., Springfield, VA. 22161, HCP/M5037-01. This is the publication on which this book, **Energy Efficient Site Design,** is based.

Ridgeway, James, **Energy Efficient Community Planning,** The J.G Press, Emmaus, Pa., 1979.

Roberts, James. **Energy, Land Use and Growth Policy: Implications for Metropolitan Washington**. Washington Metropolitan Area Council of Governments. June 1975.

Robinette, Gary O. **Plants, People and Environmental Quality**. Washington, D.C.: U.S. Department of the Interior, National Park Service. 1972.

Rogers, Tyler Stewart. **Design Guidelines for the Conservation of Energy**. Madison, Wisconsin: Department of Administration, State Bureau of Facilities Management. September 1973.

SMIC. **Inadvertent Climate Modification: Report of the Study of Man's Impact on Climate**. Cambridge, Massachusetts. M.I.T. Press. 1971.

Schiler, Marc and D. Greenberg, **The Calculation of Foliage Shading Effects in Computer Simulation of Building Energy Loads,** Proceedings: 16th Annual Design Automation Conference, San Diego, 1979, New York, Association for Computing Machinery, 1979.

Seifert, William W.; Baker, M. A.; Kettani, M. A. **Energy & Development: A Case Study**. Cambridge, Massachusetts. M.I.T. Press. 1973.

Sewell, W.R., Derrick and Harold D. Foster, **Energy Conservation Through Land Use Planning,** Environment Canada, Lands Directorate Working Paper No. 6, Ottawa, 1980, 97 pp.

Simon, Andrew L. **Energy Resources**. New York: Pergamon Press. 1975.

Simon M. **Energetic Arguments: On Site Energy**. Architectural Record. April 1976.

Smil, Vaclay. **Energy & The Environment: A Long Range Forecasting Study**. Winnipeg: Department of Geography, University of Manitoba. 1974.

Smith, Edward P., **A System for Plant Selection in Planting Design,** University of California, Berkeley, CA., 1965.

Socolow, R. H. **Energy Conservation in Housing**. Princeton, New Jersey: Center for Environmental Studies. 1974.

Sporn, Philip. **Energy in an Age of Limited Availability & Delimited Applicability**. Oxford, New York: Pergamon Press. 1976.

Steadman, Phillip. **Energy, Environment & Building**. New York: Cambridge University Press. 1975.

Stephenson, Lee, **Energy Manual for Parks: A Handbook for Interpreters and Naturalists,** National Recreation and Park Association, Arlington, VA., 1976

Stromberg, R.P. and S.O. Woodall, **Passive Solar Buildings: A compilation of of data and results,** Sandia National Laboratory, Albuquerque, N.M., 1977.

Sunset Books, **Low Maintenance Gardening,** Lane Publishing Co., Menlo Park, CA., 1974.

Stein, Jane, editor. **New Energy Technologies for Buildings**. Institutional Problems and Solutions. Cambridge, Massachusetts: Ballinger Publishing Co. 1975.

Taussig, R. T. **Commercial Energy Use**. Columbia University, Pergamon Press, Inc. 1976.

Technology + Economics, Inc., **The Energy-Wise Home Buyer,** A Guide to Selecting an Energy Efficient Home, U.S. Department of Housing and Urban Development and the U.S. Department of Energy, Washington, D.C., 1979, 60 pp.

Thirring, Hans. **Energy for Man: From Windmills to Nuclear Power**. New York: Greenwood Press. 1968.

U.S. Department of Commerce. **Making the Most of Your Energy Dollars in Home Heating & Cooling**. Washington, D.C.: National Bureau of Standards. June 1975.

U.S. Department of Energy, Community Energy Program, Community Systens Division, assisted by the Energy and Environmental Systems Division, Argonne National Laboratory, **How Do You Achieve Energy Efficient Development?,** brochure developed by the Center for Landscape Architectural Education and Research, Community Energy Program, Site and Neighborhood Case Studies, available from the U.S. Department of Energy 111.3, Washington, D.C., 20585.

U.S. Department of Energy, Community Energy Program, Community Systems Division, assisted by the Energy and Environmental Systems Division, Argonne National Laboratory, Community Systems Project Profile, **Site and Neighborhood Design for Energy Conservation,** brochure developed by the Center for Landscape Architectural Education and Research, available from the U.S. Department of Energy, Washington, D.C., 20585.

U.S. Housing and Home Finance Agency. **Application of Climatic Data to House Design**. Washington, D.C.: (Housing Research) 1954.

United Nations. **Climate and House Design**. New York: United Nations, Department of Economic-Social Affairs. 1971.

Valko, P. **The Effect of Shape & Orientation on The Radiation Impact on Buildings**. Stockholm: CIM/WMO Colloquim on Building Climatology, "Teaching the Teachers." Paper No. 33. September 1972.

Van der Meer, Wybe J. **Research and Innovation in the Building Regulatory Process**, "Avoid Tunnel Vision in Implementation of Energy Conservation Building Standards," edited by Patrick W. Cooke. Providence, Rhode Island: Proceedings of the NBS/NCBBCS Joint Conference. September 1976. Washington, D.C.: National Bureau of Standards. June 1977.

Van Dresser, P. **A Landscape for Humans**. Albuquerque, New Mexico: Biodynamic Press. 1972.

Wells, Malcolm. **Underground Design**. Cherry Hill, New Jersey: 1977.

Williams, Robert H., editor. **The Energy Conservation Papers**. A Report to the Energy Policy Project of the Ford Foundation. Cambridge, Massachusetts: Ballinger Publishing Company. 1975.

Wittmus, Waldemar A. **The Effect of Trees on Wind as Related to Drifting Snow**. University of Wisconsin, Department of Landscape Architecture. 1940.

Wolozin, Harold. **Energy and the Environment: Selected Readings**. Morristown, New Jersey: General Learning Corporation. 1974.

Yellot, John I. **Plant Engineering**. Arizona: Yellot Solar Energy Lab. April 17, 1975.

Zaelke, Durwood J. **Saving Energy in Urban Transportation**. Cambridge, Massachusetts: Ballinger Publishing Co. 1977.

PERIODICALS

AIA Journal. "AIA Asks Wide Support for Energy Conservation." August 1976.

AIA Journal. "Energy Report Focuses on Environment." January 1976.

Abbasi, S. "Environmental, Energy & Land Use Issues." **American Institute of Planners**. 1975.

An Energy and Ecological Analysis of Alternative Residential Landscapes, **Journal of Environmental Systems,** 1982, V. 11, n.3, p.271. 20-82-23994.

Architectural Forum. "Energy Retrofitting." March 1974.

Architectural Forum. "Selected List of Energy Research Organizations." July/August 1973, p. 75.

Architectural Record. "1 + 1 = 3: A New Equation for Counting a New Building's Cost." December 1975, pp. 68-83.

Architectural Record. "Princeton University Studies New Town Energy Use." August 1973.

Bacon, Edmund. "Energy and Land Use." **Urban Land**. July/August 1973.

Barnes, Edward Larrabee. "College of the Atlantic; Maine, Bar Harbor; Energy Conservation Uses." **Progressive Architecture**. January 1974, pp. 62-63.

Bates, C. G. "The Wind Break as a Farm Asset." Washington, D.C.: **Farmer's Bulletin**, No. 1405. Government Printing Office. 1944.

Beeney, Bill. "Energy House Nearing Completion: Cooperative Project on RIT." Rochester, New York: **Democrat & Chronicle**, Real Estate Section. August 7, 1977, pp. 1H & 2H.

Berkoz, Esher Balkan. "Optimum Building Shapes for Energy Conservation." **Journal of Architectural Record**. Mid-August 1974, pp. 144-148.

Better Homes and Gardens "46 Ways to Conserve Energy in Your Home." November 1973, p. 36.

Bridgers, Frank. "Energy Management is a Way of Life; Benefits from Solar Energy." **Architectural Record**. Mid-August 1974, pp. 144-148.

Brown, Vincent M. "Energy—Here Today." **Urban Land**. July/August 1973.

Building Systems Design. "Cities as Energy Systems." February/March 1976.

Building Systems Design. "Energy Conservation Through Rational Planning and Architecture." June/July 1976.

Carr, Donald E. "The Lost Art of Conservation." **Atlantic**. December 1975, pp. 59-61.

Commoner, Barry. "Alternative Approaches to the Environmental Crisis." **American Institute of Planners Journal**. May 1973.

Cooperative Extension Association of Monroe County. "Save Energy, Save Dollars." A supplement to the **Democrat & Chronicle**. October 20, 1977, pp. 12A-E.

Dallaire, Gene. "Designing Energy—Conserving Buildings." **Civil Engineering ASCE**. April 1974, pp. 54-58.

Daly, Leo A. "Energy and the Built Environment: A Gap in Current Strategies." **AIA Journal**. May 1974.

Davis, Martin. "Solar Greenhouse Residence Project ." **Journal of Architectural Education**. Vol. 30, No. 3, February 1977, pp. 21-24.

Dean, O. "Redesigning an Entire Town (and its Life Style) for Energy Conservation." **AIA Journal**. November 1975.

Deering, R. B. "Effective Use of Living Shade." **California Architect**. September 1955, pp. 10, 11-15.

Dempewolff, Richard F. "10 Worst Heat Thieves in Your Home and How to Halt Them." 1976 Reprint from **Popular Mechanics**.

Designing and Evaluating Energy Efficient Landscape Plantings, **Solar Engineering,** September 1981, v. 6, no. 9, p. 14, 20-82-22483.

Doxiadis, C. A. "Energy and Human Settlements." **Ekistics**, November 1968.

Dubin, Fred S. "Energy for Architecture." **Architecture**. July 1973.

Dubin, Fred S. "Energy Conservation Needs New Architecture and Engineering." **Public Power**. March/April 1972.

Everson, G. F.; Newbauer, L.W.; Deering R. B. "Environment Influence on Orientation and House Design to Improve Living Comfort." **Journal of Home Economics**, Vol. 48, No. 3. March 1956, pp. 161-167.

Fuchs, John. "A Realistic Approach to Conserving Energy." **Progressive Architecture**. September 1973.

Geddes, Robert. "Nature of the Built Environment; Relationship of Landscape to Building." **Progressive Architecture**. June 1974, pp. 72-81.

Gold, Seymour M. "Recreation Planning for Energy Conservation." **Parks & Recreation**. September 1977, pp. 61-63, 83.

Guidelines for Climatic Efficient Residential Planning, **Urban Land,** May 1981, v. 40, no. 5, p. 16. 20-81-24996.

Hemphill, Marion L. **Environmental Comment**. "The Portland Energy Conservation Project." July 1977.

House & Home. "Energy Savings by Upgrading Specs, Orientation, Landscaping." April 1975, pp. 59-63.

Howland, Joseph E. "How Privacy Can Increase Your Living Space and Improve Your Climate." **House Beautiful**. June 1950, pp. 101,102.

Johnson, Timothy C. "Performance of Passively Heated Buildings." **Journal of Architectural Education**, Vol. 30, No. 3, February 1977, pp. 16-20.

Keyes, Dale L. "Energy and Land Use: An Instrument of U.S. Conservation Policy." **Energy Policy**. September 1976.

Knowles, Ralph. "Solar Energy, Building and the Law." **Journal of Architectural Education**, Vol. 30, No. 3, February 1977, pp. 68-72.

Landscaping for Cooling, **Alternative Sources of Energy,** July-August, 1982, n. 56, p. 8, 20-82-24982.

Langeweische, Wolfgang. "Can You Control the Wind?" **House Beautiful.** June 1950, pp. 88, 91, 97, 204.

Langeweische, Wolfgang. "How to Control the Sun." **House Beautiful.** March 1950, p. 131.

Langeweische, Wolfgang. "How to Fix Your Private Climate." **House Beautiful.** October 1949, pp. 108, 147, 150, 152, 192.

Langeweische, Wolfgang. "How to Live Comfortably in the Southwest Desert." **House Beautiful.** April 1950, pp. 205-207.

Langeweische, Wolfgang. "How to Pick Your Private Climate." **House Beautiful.** October 1949, pp. 147, 148.

Langeweische, Wolfgang, "There's a Gold Mine Under Your House." **House Beautiful.** August 1950. pp. 91,92, 93, 125.

Langeweische, Wolfgang. "Wind Control in Hot Weather." **House Beautiful.** June 1950, pp. 124, 165.

Leighton, Gerald S. "Evolution of an Idea: HUD Takes Total Energy Concept One Step Beyond." Reprinted from the January 1973 issue of **Aircondition & Refrigeration Business.**

Lichtenberg, A. J. & Schipper, L. "Efficient Energy Use & Well Being: The Swedish Example." **Science.** April 8, 1977.

Marshall, Harold E. & Ruegg, Rosalie T. "Energy Conservation Through Life-Cycle Costing." **Journal of Architectural Education,** Vol. XXX, No. 3, February 1977, pp. 42-51.

McCallum, D. L. (Reply to "Alternative Approaches to the Environmental Crisis") **American Institute of Planners Journal.** March 1974.

McGregor, Gloria Shepard. "Davis, California, Implements Energy Building Code." **Practicing Planner.** February 1976. pp. 24-6, 31.

McGregor, Gloria Shepard. "Davis, California: A Pace-Setting Energy Conservation City." **Environmental Comment.** July 1977.

Moorcraft, C. "Solar Energy in Housing." **Architectural Design.** 43: 634.

National Association of Home Builders. **Land Development Manual.** Washington, D.C.: NAHB, c. 1974.

Nelson, W.R., Landscaping Beautifies Buildings and Conserves Energy, **American Nurseryman,** 150:16+, September 1, 1979.

Neubauer, L. W. "Orientation and Insulation: Model Versus Prototype." **Transactions of American Society of Agricultural Engineers.** Vol. 15, November 9, 1972, pp. 707-709.

Neubauer, L. W. & Cramer, R. D. "Shading Devices to Limit Solar Heat Gain But Increase Cold Sky Radiation." **Transaction of American Society of Agricultural Engineers.** Vol. 8, November 4, 1956, pp. 470-472, 475.

New York Times. "H. R. Senate Committee Approves Energy Conservation Bills." July 31, 1976, 24:1.

New York Times. "School Opens in Reston, Virginia, with Energy Saving Devices and Solar Collectors Paid for with $665,000 Grant From Saudi Arabian Government." Vol. 4, February 27, 1977, 9:1.

Odell, P. R. "Energy Alternatives: Planning Implications." **Town and Country Planning**. January 1975.

O'Donnell, Robert M. & Parker, James E. "Large-Scale Development: A Breeder for Energy Conservation." **Environmental Comment**. July 1977, pp. 3-5.

Odum, H. T. & Peterson, L. L. "Relationships of Energy and Complexity in Planning." **Architectural Design**. Vol. 42, 1972, pp. 624-629.

O'Sul, P. E. "Integrated Environmental Design of Buildings. **Ekistics**. May 1972, p. 355.

Outdoor Recreation Action. "Energy Crisis and Recreation." Spring 1974.

Pauly, David & Cook, William J. "How to Save Energy." **Newsweek**. April 18, 1977, pp. 70-74.

Pitts, D.G. and W. Gould, Jr. Designing Home Landscapes for Energy Conservation, **American Nurseryman,** 152:11+, November 1, 1980.

Pitts, D.G. and others, A New Approach to Conservation: Energy Consumed by the Landscape, **American Nurseryman,** 155:60-65, February 1, 1982,

Professional Engineer. "Energy Conservation Through Design." October 1973, pp. 21-24.

Progressive Architecture "Energy Computer Programs and Other Services." May 1975, pp. 102-104.

Rancer, Michael D. "Energy Conservation in New Residential Buildings." **Municipal Management Innovation Series**, No. 17. April 1977.

Science. "Energy Conservation in New Housing Design." June 26, 1976.

Simon, M. "Energetic Arguments: On Site Energy." **Architectural Record**. April 1976.

Smay, V. Elaine. "Underground Houses." **Popular Science**. April 1977, pp. 84-89, 155.

Smith, P., What's A Tree Worth?, Reducing Home Energy Demands, **Family Handyman,** 29: 38-39, May 1979.

Steadman, Philip. "Energy and Patterns of Land Use.: **Journal of Architectural Education**, Vol. XXX, No. 3, February 1977, pp. 36-40.

Trechel, Heinz R. "Research in Energy Conservation." **Journal of Architectural Education**, Vol. XXX, No. 3, February 1977, pp. 31-33.

Sunset Magazine. "Wind Test—Which Fence is Best." **How to Build Fences and Gates**. Lane Publishing Co.

Temperate-Zone Landscape Designs for Home Heating and Cooling, **Horticulture,** May 1981, v. 59, no. 5, p. 18, 20-81-24995.

Tree Plantings That Save Energy, **Organic Gardening,** October 1980, v. 27, no. 10, p. 68, 20-81-20987.

Watson, Donald, ed. "Energy and Architecture." **Journal of Architectural Education**. February 1977.

Watson, Donald. "Energy Conservation in Architecture. Part 1: Adaptive Design to Climate," and "Energy Conservation in Architecture. Part 2: Alternative Energy Sources." **Connecticut Architect**. March-April and May-June 1974.

Wells, Malcom B. "Conservation Architecture. **Architectural and Engineering News**. September 1969, p. 70.

Wheeler, Alan A. "Mission Viejo, California: A Pace-Setting Energy Conservation Community." **Environmental Comment**. July 1977, pp. 5-7.

Wilke, Douglas A. & Fuller, David R. "Highly Energy Efficient—Wilton Wastewater Treatment Plant." **Civil Engineering ASCE**, Vol. 46, May 1976, p. 70.

Williams, G. S., Your Landscaping Can Conserve Energy, **Flower and Garden**, v. 24, p. 48-49, September 1980.

Wright, Henry. "How to Put a Harness on the Sun." **House Beautiful**. October 1949. pp. 153, 158, 185.

Wright, James R.; Aehenbach, Paul R. "Energy Conservation in Buildings." **Scientific American** Roundtable. January 1974.

REPORTS

AIA Research Corporation. "Decision Making in the Building Process." August 1976.

American Society of Landscape Architects. "Site Specific Solutions for Energy Development and Conservation."

Argonne National Lab. "Regional Energy Modeling: An Evaluation of Alternative Approaches." Argonne, Illinois: June 1975.

Associates, Inc. "In the Bank . . . or up the Chimney." Cambridge, Massachusetts: April 1975.

Association of Bay Area Governments. "Energy Needs—A New Criteria for Planning." February 1974.

Association of Physical Plant Administrators of Universities and Colleges, Energy Project Office. "Feasibility Study on the Impact of Agencies and Codes on University and College Energy Use." (Prepared for Energy Research and Development Administration, Division of Buildings and Community Systems.) Springfield, Virginia: National Technical Information Service, U.S. Department of Commerce. March 1977.

Auerbach, S. I. "Ecology's 20-Year Space Program." **ORNL Review**. Vol. 8, No. 1, Winter 1975, p. 21.

Bainbridge, David; Hammond, J.; Hunt, Marshall. "Community Design for Energy Conservation." Sacramento County Energy Planning and Conservation Council/Living Systems. 1976.

Bainbridge, David; Kopper, B.; Hammond, J.; Hunt, Marshall. "The Davis Energy Conservation Report: Practical Use of the Sun." Living Systems. 1977.

Bather, Ringrose, Wolsfeld, Inc. Minnesota Campus Long Range Development Planning Framework. University of Minnesota, Office of Physical Planning; Roger DuToit, Architects and Planners. January 1976.

Booz Allen and Hamilton, Inc. "Senitruity of the Leisure Recreation Industry to the Energy Crisis." Federal Energy Office. January 18, 1974.

Botha, Philip R. "Building and Land: The Nature of the Relationship." Berkeley, California: 1959.

Braiterman, Marta. "Energy and Land Use: An Assessment of Selected Landscape Resources For Use in the Metropolitan Landscape Research Planning Model (Method). University of Massachusetts (Thesis). October 1977.

Broome County Department of Planning. "Environmental Considerations for the Site Plan Review Process of Land-Use Planning." Binghamton, New York.

California, State of: Division of the Department of General Services. "A Competition for an Energy Efficient Office Building." Sponsored by the California Energy Resources Conservation and Development Commission and the Office of the State Architect.

Carroll, T. O., et al. "A Systems View of Energy and Land Use." Upton, New York: Brookhaven National Lab. Sponsored by FEA. October 1975.

Cerritos, City of. Ordinance No. 475. California, County of Los Angeles: June 5, 1974. Agnes Hickey, City Clerk.

Cohen, Allen S. & Costello, Kenneth W. "Regional Energy Modeling: An Evaluation of Alternative Approaches," Argonne National Lab. Argonne, Illinois. June 1975.

Cohn, Sidney. **Research and Innovation in the Building Regulatory Process**. "Effectiveness of U.S. Municipal Design Review Programs —Preliminary Findings." Edited by Patrick W. Cooke. Proceedings of the First NBS/NCBBCS Joint Conference Held in Providence, Rhode Island on September 21-22, 1976. Washington, D.C.: National Bureau of Standards, June 1977.

A Comparative Analysis of the Role of Various Elements in Passive Cooling in Warm, Humid Climates, presented at ISES-AS/et.al, Passive & Hybrid Cooling Intl. Conf., Miami Beach, Nov. 6-16, 1981, p.365, 10-82-21207.

Conserving Energy With Landscape Planting, presented at Information Transfer, Inc./et.al. Technology for Energy Conservation Conf., Tucson, Jan. 23-25, 1979, p. 151, 20-80-21477.

Cost Effectiveness of Landscaping for Energy Savings: A Case Study, presented at DOE/ et.al. 4th Natl. Passive Solar Conf., Kansas City, Oct. 3-5, 1979, p. 476. 10-80-22860.

Directory of Federal Agencies (Engaged in Energy Related Activities). Washington, D.C.: Office of Intergovernmental Regional and Special Programs, FEA Item No. 439-A-1; FE1.2:F31. October 1975.

Directory of State Government Energy Related Agencies. Washington, D.C.: National Energy Information Center, FEA 1C-75/515. Item 434A-1; FE 1.d2: St.2. September 1975.

Dodds, Douglas. "Energy Conservation: An Introductory Guide for Local Government." Professional Report. Berkeley, California: University of California, Department of City and Regional Planning. June 1977.

The Effects of Orientation and Shading from Trees on the Inside and Outside Temperatures of Model Homes, presented at ISES-AS/et.al., Passive & Hybrid Cooling Intl. Conf., Miami Beach, Nov. 6-16, 1981, p. 369, 10-82-21208.

The Effectiveness of an Evergreen Windbreak for Reducing Residential Energy Consumption, ASHRAE Trans., 1979, v. 85, no. 2, p.428, 20-80-25472.

Energy Conservation and Community Development, Sen. Comm. Banking Housing & Urban Hearings, 96 Cong., 1 Dec. 13, 1979. 17-81-22396.

Energy Conservation Landscaping as a Passive Solar System, presented at DOE/et.al., 4th Natl. Passibe Solar Conf., Kansas City, Oct. 3-5, 1979, p. 471. 10-80-22859.

Energy Conservation Landscaping as a Passive Solar System, presented at DOE/et.al., 4th Natl. Passive Solar Conf., Kansas City, Oct. 3-5, 1979, p. 476, 10-80-22860.

Educational Facilities Laboratories. "Energy and Educational Facilities: Costs and Conservation." Springfield, Virginia: National Technical Information Service. April 1977.

Edwards and Schafer. "Relationships Between Transportation Energy Consumption and Urban Structure Results of Simulation Studies." Northwestern U.S. Department of Civil Engineering. January 1975.

Energy Conservation in Urban Development, Swedish Council for Building Research, Report D9-1980, 1980, 20-81-24-993.

Energy Research and Development Administration. "Building and Community Systems." Washington, D.C.: Office of Conservation, Division of Buildings and Community Systems. September 30, 1976.

Energy Research and Development Administration. "Model Code for Energy Conservation in New Building Construction." Draft Report. Energy Conservation Program. June 1977.

Energy Research and Development Administration. "A National Plan for Energy Research, Development and Demonstration: Creating Energy Choices for the Future." Vol. 2 Program Implementation. Washington, D.C.: June 30, 1976.

Engineering Foundation Conference. "Energy Conservation Through Efficiency Energy Utilities." Henniker, New Hampshire: New England College. August 18-24, 1973.

Erly, Duncan and David Mosena, Energy Conserving Development Regulations: Current Practice, American Planning Association, P;A.S. Report 352, Chicago, 1980, 58 pp.

Federal Energy Administration. "Energy Conservation Site Visit Report: Toward More Effective Energy Management." Washington, D.C.: Conservation Paper Number 38. April 1976.

Fostel, Henry Fabian. "Energy Conservation Through Management of the Microclimate: Impact of Land Development on the Natural Microclimatic Heat and Water Balances.: Cambridge, Massachusetts: Graduate School of Design; Department of Landscape Architects and Landscape Architectural Research Office. March 1975.

Fraker, Harrison & Schoncke, Elizabeth. "Energy Husbandry in Housing: An Analysis of the Development Process in a Residential Community." Twin Rivers, New Jersey: Princeton University. December 1973.

Friedman, Charlotte F. "Building Standards and Energy Conservation in Public School Buildings: An Overview of Three States: Colorado, Rhode Island, Virginia." Arlington, Virginia: Office of Governmental Relations. ·American Association of School Administrators. April 1977.

Garvey, Gerald. "Energy, Ecology, Economy. New York: Norton. 1972.

Goldberg, Philip. "Planning with Energy." Philadelphia, Pennsylvania: Rahenkamp, Sachs, Wells and Associates, Inc. January 1975.

Gray, R. L. "The Impact of Solar Energy on Architecture." University of Oregon. 1974.

Greeley, Richard S. "Energy Use and Climate: Possible Effects of Using Solar Energy Instead of Stored Energy." Washington, D.C.: National Science Foundation. April 1975.

HUD. "Solar Heating and Cooling Demonstration Program: A Descriptive Summary of HUD Cycle 2 Solar Residential Projects." U.S. Development and Research, Division of Energy, Building Technology and Standards in cooperation with Energy Research and Development Administration. Fall 1976.

Hammond, Jonathan; Hunt, Marshall; Cramer, Richard; Neubauer, Loren. **A Strategy for Energy Conservation**: Proposed Energy Conservation and Solar Utilization Ordinance for the City of Davis, California: The City of Davis, California. 1974, pp. 14, 41. Cited p. 53.

Harrington, Winston. "Energy Conservation: A New Function for Local Government." University of North Carolina. December 1976.

Harvey, D. G. "Industry Conservation." Program Approved Document: Executive Summary. Prepared for Energy Research and Development Administration. December 19, 1975.

Holloway, Dennis, Project Director. "Winona: Towards an Energy Conserving Community." Minneapolis, Minnesota: University of Minnesota, School of Architecture and Landscape Architecture. c 1975.

Ilse, J. Master's Thesis. Cited in **Solar News and Views** June 1976, p. 4.

Johnson, G. R.; Shaner, W. W.; Hall, D. E. & Sheng, T. **Energy Conservation by Passive Solar Design Factors**. Fort Collins, Colorado: Unpublished paper, Department of Mechanical Engineers, Colorado State University. 1975.

Kaminsky, Jacob. "Environmental Characteristics Planning; An Alternative Approach to Physical Planning." Baltimore, Maryland: Regional Planning Council. July 1972.

Keyes, Dales L. "Urban Form and Energy Use." Urban Institute. September 1975.

Koppel, William. "Energy Use and Conservation in Davis Households." Davis, California: For M.S. Ecology Degree, University of California 1974.

Knight, Robert L. & Menchik, Mark D. "Residential Environmental Attitudes and Preferences: Report of a Questionnaire Survey." IES Report 24. Madison, Wisconsin: University of Wisconsin, Institute for Environmental Studies, Center for Human Systems. October 1974.

Kurilko, George N. "Design Practices to Conserve Energy." Site Specific Solutions for Energy Development and Conservation, by the ASLA, Colorado Chapter.

Landscape as Energy and Environmental Conservator in the Arid Regions-Saudi Arabia, presented at ISES-AE/et.al., Passive & Hybrid Cooling Intl. Conf., Miami Beach, Nov. 6-16, 1981, p. 360, 10-82-21206.

Landscape for Passive Solar Cooling, presented at ISES-AE/et.al. Passive & Hybrid Cooling Intl. Conf., Miami Beach, Nov. 6-16, 1981, p. 360, 10-82-21206.

Levine and McCann. "TIES. Total Integrated Energy System Feasibility Analysis for the Downtown Redevelopment Project. Pasadena, California." Pasadena, California: Levine and McCann, A Division of Syska & Hennessy & Genge, Community Consultants. Prepared by the City of Pasadena, Pasadena Redevelopment Agency for Energy Research and Development Administration. April 1977.

Living Systems. "Energy Conservation Building Code Workbook." Winters, California: Living Systems and City of Davis, California, July 1976.

Living Systems. "Davis Energy Conservation Report: Practical Use of the Sun." Winters, California: Living Systems. April 1977.

Living Systems. "Indio, California: Energy Conservation Project." Final Draft Report. Winters, California: Living Systems. March 16, 1977.

Longley Research Center. "Technical Support Package." Hampton, Virginia: Longley Research Center for NASA Tech. Brief. LAR-12/34 NASA Technology Utilization House. Winter 1976.

Mattingly F. & Peters, Eugene F. "R-20 Wind and Trees—Air Infiltration Effects on Energy in Housing." Princeton, New Jersey: Center for Environmental Studies, Princeton, University. May 1975.

McGregor, Gloria S. Project Manager. "Davis Retrofit Manual." City of Davis, California. June 1, 1976.

McGregor, Gloria S. Project Manager. "Energy Conservation Advisory Program." Davis, California: City Council. HUD Innovative Project B75-S1-06-001. June 1976.

Mills, Jon L. et al., "Energy: The Power of the States." Gainesville, Florida: University of Florida, Center for Governmental Responsibility, Holland Law Center. October 1975.

Mission Viejo Company. "Minimum Energy Dwelling (MED)." (Brochure) Mission Viejo, California: Research sponsored by Energy Research and Development Administration, Southern California Gas Company, Los Angeles, California, and the Mission Viejo Company. 1975.

Moreland, Frank L. Editor. "Alternatives in Energy Conservation. The Use of Earth Covered Buildings. Fort Worth, Texas: Proceedings of a conference. July 9-12, 1975.

Municipal Attorney. "Conservation of Energy in Chattanooga." Chattanooga, Tennessee: January 1975.

National Research Council. "Weather and the Building Industry." Research Conference. Building Research and Advisory Board, Confidential Report No. 1. January 11-12, 1950.

National Weather Records Center. "Selective Guide to Climatic Data Sources." Asheville, North Carolina: Staff, National Weather Center. Documentation No. 4.11. 1969.

OECD Publications. "Better Towns with Less Traffic." Washington, D.C.: April 1975.

Pacific Gas and Electric Co. "Energy Economics in Building Design." Symposium II—Proceedings of a conference held October 19, 1974.

Passive Cooling Design Approaches, a Review of Applicability for Hot Humid Climates, presented at DOE/et.al., 4th Natl. Solar Conf., Kansas City, Oct. 3-5, 1979, p. 517, 10-80-22866.

Pope, Evan and Robbins. "Site Energy Handbook." Washington, D.C.: Item 1051-C, Biblio. A1-5. U.S. Energy Research and Development Administration, Division of Building and Community Systems. 1976.

Precision Landscaping for Energy Conservation, presented at Information Transfer, Inc./et.al., Technology for Energy Conservation Conf, Tucson, Jan. 23-25, 1979, p. 151, 20-80-21477.

Princeton University Centre for Environmental Studies. "Energy Husbandry in the Home Building Industry—A Look at Planned Unit Development." March 28, 1974.

Public Buildings Service. "Energy Conservation Guidelines for Existing Office Buildings." February 1975.

Public Planners, Administration and Regulations. "Institutional Factors Influencing the Acceptance of Community Energy Systems and Energy Efficient Community Design." ASPO. 1976.

Read, Ralph A. "Tree Windbreaks for the Central Plains." **U.S. Forest Service Agriculture Handbook**, No. 250. February 1964.

Reiter, Elmar R. et al. "The Effects of Atmospheric Variability on Energy Utilization and Conservation." Fort Collins, Colorado: Department of Atmospheric Science, Colorado State University. (Final Report of Research Conducted between July 1, 1975 and October 31, 1976.) No. 5. November 1976.

Romanos, Michael C. "Community Planning for Energy Conservation. A Scenario for ERDA's Community Design Research and Development and Demonstrations." Springfield, Virginia: National Technical Information Service. June 1977.

Salter, Richard G. & Morris, Dean N. "Energy Conservation in Public and Commercial Building." The Rand Paper Series. Santa Monica, California: The Rand Corporation. October 1973.

Schiler, Marc, Foliage Effects on Computer Simulation of Building Energy Load Calculations, Thesis, Cornell University, Ithaca, N.Y., 1979.

Schiler, Marc and D Greenberg, The Calculations of Translucent and Opaque Shadow Effects in Building Energy Simulation, Proceedings: CAD 80, the 4th International Conference on Computers in Architecture and Engineering, Brighton 1980, Guildford, Surrey, IPC Science and Technology Ltd., 1980.

Sewell, W. R, Derrick and Harold Foster, Analysis of the United States Experience in Modifying Land Use to Conserve Energy, Environment Canada, Lands Directorate, Working Paper No. 2, 1980, Ottawa, 192 pp.

Shenandoah Development, Inc. Energy Conserving Site Design Case Study-Shenandoah, GA., U.S. Department of Energy, Washington, D C., 1980., available from the National Technical Information Service, (N.T. I.S) 5285 Port Royal Rd., Springfield, VA. 22161.

United States Department of Housing and Urban Development. "Solar Heating and Cooling Demonstration Program." Washington, D.C.: Office of Policy Development and Research. July 1976.

United States Energy Research and Development Administration. "Energy and Educational Facilities: Costs and Conservation." Oak Ridge, Tennessee: April 1977.

United States Energy Research and Development Administration. "A Feasibility Study on the Impact of Agencies and Codes on University and College Energy Use." Oak Ridge, Tennessee: USERDA Technical Information Center. March 1977.

Using Vegetation to Cool Small Structures, presented at ISES-AE/et. al. 1981 Annual Conf., Philadelphia, May, 26-30, 1981, v. 2, p. 905, 10-81-23767.

Villecco. "Energy Conservation Opportunities in Building Design." Washington, D.C.: AIA Research Corporation. October 1973.

Walters, George S. "Ground Surface Treatments in Landscaping Maintenance Problems." Berkeley, California: University of California. 1954.

Wanben, D. & Rugg, W. "Energy Needs—A New Criterion for Planning." ADAG. 1974.

White, R. F. "Effects of Landscape Development and Natural Ventilation." Austin, Texas: Texas Engineering Experiment Station. 1945.